5

£4.50

WATERWAYS AND WATER LIFE OF GREAT BRITAIN

Heather Angel & Pat Wolseley

WATERWAYS AND WATER LIFE OF GREAT BRITAIN

Photography by Heather Angel

PEERAGE BOOKS

First published in 1982 by
Michael Joseph Limited under the title
The Family Water Naturalist

This edition published in 1986 by
Peerage Books
59 Grosvenor Street
London W1

This book was created and produced by
Roxby Water Naturalist Limited
98 Clapham Common Northside
London SW4

ISBN 1 85052 055 0

Printed in Spain

CONTENTS

ACKNOWLEDGEMENTS

Pat Wolseley would like to acknowledge her debt to the Field Studies Council, in particular to Charles Sinker, the Director, for his inspiration and contributions in the early stages. The authors would also like to thank him for contributing the copy for 'Troubled Waters'; and Adrian Bayley for 'Swimmers and Clingers', 'Burrowers and Hiders', 'Animal Architects', 'Prey and Predators' and 'The Stream at Night'. They both also contributed greatly to discussions on the practicalities of various freshwater projects.

The authors would also like to thank Dick Phillips for arranging their joint trip to Iceland.

Photography and artwork
The photographs are by Heather Angel, with the exception of those in the following list. Both authors extend grateful thanks to all these sources of photography and artwork.
Aerofilms: Spurn Head (p. 118); Avon Rubber Co. Ltd: Thames in flood (p. 97); H. Boyd, The Wildfowl Trust: duck decoy (p. 178–9); British Trust for Conservation Volunteers: volunteers at work (p. 56); Cambridge University Collection (copyright reserved): land reclamation Holland (p. 176), Norfolk Broads (p. 177); Central Electricity Generating Board (North-West Region): Connah's Quay Power Station (p. 116); John Clegg: water spider with air bell (p. 77); Clyde Surveys Ltd: infra-red scan of Rotterdam dockside (p. 116), tanker discharging oil (p. 116); Professor John Coles, Somerset Levels Project: references (pp. 52–3); B. A. Crosby, The Wildfowl Trust: Ouse washes (p. 176); Claire Dalby: her excellent drawings on pp. 30–1, 48–9, 54, 68–9; E.R.D. Publications Ltd, Exmouth: Lynmouth disaster (p. 12–13); Evans Bros. Ltd: basis of glaciation map (p. 16); Faber and Faber Ltd: basis of reconstruction of Rhine (p.94);

Greater London Council (by courtesy): Thames Barrier (p. 97); Handford Photography, Croydon: Kingsnorth Power Station (p. 117); Bert Hawkes, Aston University: basis of pollution diagram (pp. 84–5); Valerie Hill: drawings; H.M.S.O: Claish Moss (p. 53); Institute of Oceanographic Sciences: sonograph of trawl marks (p. 172); R. E. Jones: artwork of waders' bill lengths (p. 110); Dr B. E. Juniper and D. Kerr, Botany Schools, Oxford: SEMs of Sphagnum cells (p. 53); Dr B. E. Juniper and G. Wakely: SEMs (pp. 58–9); Aziz Khan: diagrams; Kestins: card of Pulpit Rock (p. 155); Port of London Authority: driftwood on River Thames (p. 96); K. A. Pyefinch: basic data for graph (p. 151); Brian Rogers/ Biofotos: ringed plover eggs (p. 130), black-tailed godwit (p. 56); Scientific American: basis of freshwater volumes diagram (p. 10); Seaphot Ltd: diver (p. 9); Soames Summerhays/Biofotos: iceberg (p. 11), Galapagos fur seal (p. 9), grey whale blowing (p. 168), oil rig (p. 171); Space Frontiers: Earth from space (p. 10–11); Charlotte Styles: drawings; Alan Suttie: majority of the excellent drawings and diagrams; Jack Terhune/Biofotos: harp seal pup (p. 169); Thames Water Authority: sewer (p. 98); John Woodcock: diagrams.

Information
The authors would like to thank the following individuals and organizations who kindly provided information:
Heather Angel:
Anglian Water Authority; Dr Roger Bamber; Brighton Corporation; British Trust for Conservation Volunteers; Central Electricity Generating Board, Press Office; Dr Robin Crump, Warden, Orielton Field Centre; Dr Pamela Harrison; Dr Andrew Heaton, Warden, Sevenoaks Experimental Wildfowl Reserve; Dr Luc Hoffman, Director, Station Biologique de la tour de Valat; Hydro Electricity Board,

Edinburgh; The International Waterfowl Research Bureau; Institute of Oceanographic Sciences; Ministry of Agriculture, Fisheries and Food; National Trust; National Water Council; Nature Conservancy Council; Preston Montford Field Centre; Royal Society for the Prevention of Cruelty to Animals; Royal Society for the Protection of Birds; Department of Agriculture and Fisheries for Scotland; Dr T. L. Shaw; Brian Stott; Dr Arthur Stride; Water Research Council, The International Service on Toxicity and Biodegradability; The Wildfowl Trust; World Wildlife Fund.
Pat Wolseley:
Professor John Coles; Dr Sylvia Haslam; Heather Howcroft; David Job; Dr Brian John; Dr Colin Reynolds.

Assistance with photography
The following people kindly helped Heather Angel to obtain photographs: Adrian Bayley; Mr. Bennett, Warden, Wicken Fen; P. H. Coate; Dr Robin Crump; Ken Davidson; Kim Debenham; Dr Andrew Heaton; John Hughes; Doug Hulyer; Peter Jones; Dick Phillips; Kevin Simmonds; Mr R. Stringer; Paul Voddon. Special thanks go to Rob Davage and Alistair Maxwell for producing the black and white prints and to Express Design Service Ltd for processing the colour transparencies.

General assistance
Heather Angel would especially like to thank Mary Stafford Smith for researching information; Liz Holder, her field assistant; Pam Blaber, Julie Burchett, Heather McFadyen and Kate West for typing; Jan McLachlan for checking galleys and designed spreads; Peter Hobson for photographic assistance; and Martin Angel for his constant involvement and support. Pat Wolseley would especially like to thank Sue Rolfe for typing copy.

▶An outwash plain in southern Iceland showing braided streams carrying glacial melt water.

THE WATER MEDIUM

A continual supply of water is vital to all forms of life on this planet. The water cycle constantly renews supplies of fresh water and on its journey creates many habitats for plant and animal communities.

Water: the basis for life

Ever since the earth cooled and the clouds of steam condensed on the surface, water has been flowing in an everlasting cycle of evaporation and precipitation. Once the water arrives on the earth's surface it begins its downhill passage to the sea, or is recycled through evaporation. It may be temporarily delayed in freshwater systems such as ponds and lakes, or in the bodies of plants and animals, or in the jungle-land of pipes and sewers maintained by man, but sooner or later it returns to the sea. Water is a vital part of all life on earth. It flows in our veins and in the sap of trees, as well as in our streams and rivers. A river entering the sea carries along its course the natural waste from plant and animal communities, together with domestic and factory waste from human settlements. Man in the modern world uses water far in excess of his biological requirements—on average 120 litres of water per person per day, much more being required for industrial and cooling purposes. How does the earth maintain a supply of fresh water, and what are the properties of water that make it so important to life?

Water as a chemical

Each water molecule is made up of one oxygen atom and two hydrogen atoms. These atoms exist in water as two charged ions. One hydrogen atom is a positively charged hydrogen ion (acid); the other has combined with the oxygen atom to produce a negatively charged hydroxyl ion (alkali). Water has remarkable dissolving powers, owing to the ability of these charged ions to combine in the same way with those of other chemicals added to it, that is, the positive and negative charges attract each other. This property of water is used everywhere in natural living processes, as well as by man in the kitchen and in the most advanced technology. Sometimes after the reshuffle of ions there is an excess of hydrogen ions, making the solution acid (like vinegar or lemon juice), or an

▼The water cycle. Freshwater, precipitated P as rain on the surface of the earth, collects into streams and rivers, or in the water table underground, on its way to the sea. Evaporation from all surfaces E, and transpiration from plant life T, renew the cycle. Man collects water in reservoirs or from the water table for domestic and industrial use and returns the water and waste products to the cycle.

▲Sunrise over Rio de Vigo, Spain, renews the cycle of evaporation of water from the sea's surface and precipitation of fresh water from the clouds.

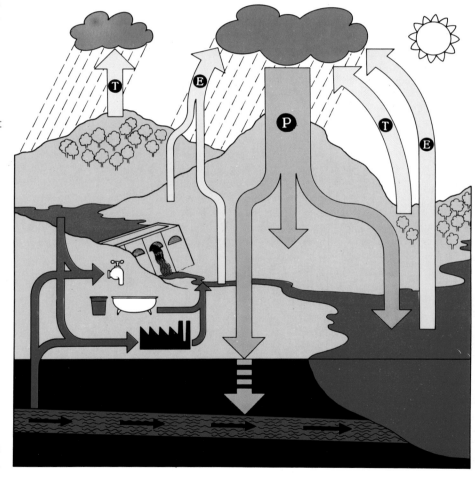

excess of hydroxyl ions, making the solution alkaline (like baking powder or caustic soda). This is measured as pH, a low pH indicating an acid solution and a high pH an alkaline one. As water passes over the surface of the earth, it dissolves substances that change its pH—so that water passing over limestone or chalk becomes more alkaline and water passing over hard silica rich rocks becomes more acid. The right balance between the pH of the plant or animal cells and the surrounding water must be maintained and, as pH may vary considerably in freshwater environments, it plays an important part in the distribution of plants and animals. However, in the sea the high concentration of salts maintains the pH at around 8, that is, slightly alkaline.

Water as a physical medium

Water exists in three phases: solid (ice), liquid (water), and vapour (steam). Moving from one phase to another

▲A diver equipped with everything that he needs to explore the shallow seas, including air tanks, a rubber suit for insulation, and flippers to extend the surface area of his feet.

▼A Galapagos fur seal *Arctocephalus galapagoensis*, whose streamlined body and short, paddle-like limbs give it all the agility and speed needed to catch the fish that are its diet.

▲There are seven pairs of gills along the abdomen of this alderfly larva *Sialis lutaria*, extending the surface area over which absorption of oxygen from the water can take place.

requires (or releases) large amounts of energy. Liquid water is denser than ice, so that ice floats on top of water. When water is warmed, it becomes less dense and warm water floats on top of cold.

Water is effective as a mechanical support for delicate organisms living in it. The mutual attraction of water molecules is also important in generating surface tension, the force which pulls a drop of water into a sphere, and which causes the 'skin' on the surface of water.

Water and life

A living cell contains over 75 per cent water within the cell wall, behaving rather like a plastic bag full of water, quite flexible but surprisingly robust. Water transports food, oxygen and waste products within the cell. It acts as a general solvent in which many of the vital chemical processes take place.

Life began in water. The problem of living in water is in keeping the right 'brew' of solutions inside while still being able both to take up oxygen and salts from the surrounding medium and to pass out waste products such as carbon dioxide and ammonia. There are two essential requisites for life that are often more limited in water than in air: oxygen and light. Oxygen is used in respiration by nearly all living organisms and is only slightly soluble in cold water, less so in warm water. Light is essential for the production of carbohydrates by photosynthesis in the plant world, but it cannot penetrate far into deep or cloudy water.

The water cycle

Water passes through three main phases in the cycle: as fresh water on the continents, as sea water in the oceans, and as water vapour in the atmosphere. The continual recycling of water requires energy on a vast scale to carry the water to the top of the cycle—the atmosphere. The sun's rays provide this energy to evaporate pure water from the surface of land and sea, leaving the impurities behind. Water vapour is carried in the atmosphere until it reaches cooler areas where it condenses as cloud or mist, and is finally precipitated as rain, hail or snow. Evaporation occurs from all water bodies and other surfaces but is most efficient over tropical seas, when there are no clouds to interfere with the strength of the sun's rays. Animals and plants also give off water vapour, particularly the green plants which constantly lose water from their leaves. Thus a tropical forest, with its enormous leaf area, makes a considerable contribution towards the recycling of fresh water; but this is still a mere drop compared with that evaporated from the oceans.

Global water

Water plays a key role in buffering the earth from extremes of temperature, so making it an easier place in which to live. The ocean absorbs and redistributes the heat from the sun, acting like an immense night storage heater, and so moderating and regulating the climate all over the globe. It also acts as a sink for naturally occurring compounds such as carbon dioxide, as well as a sink for much of man's domestic and industrial effluents.

PROJECTS

1 How much water do you drink in a day, and pass out in a day? Does it vary from winter to summer, or with smaller or larger members of your family? Where does your body lose water from? Can you work out how much water your family uses in a day, and in a week, for cooking, washing and bathing, for waste disposal and in the washing machine, by calculating the volume of the containers used? How much, on average, does each person use? Does it come anywhere near, for example, the British national average of 120 litres per person per day?

2 Obtain some pH paper that measures a wide range of pHs What pH is tap water, rain water, local stream water? Take several jars, each half full of warm water, take a pH reading, then try adding different household chemicals and substances to each. Record which substances do or do not dissolve, and those which change the pH of the water. You could try vinegar, sugar, baking powder, flour, and salt.

3 Observe how different shapes and angles affect speed in water. Using a mechanically or battery powered toy fish or boat in a bath of water, on which you have marked a 50-50cm interval with a wax crayon, note down the time it takes to travel 50 cm. Now change the shape of the toy (but not the weight) by gluing paired projections cut from a plastic bottle at different positions and angles along the toy. Repeat the experiment. How do your alterations affect the speed and direction of the toy?

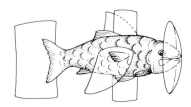

FRESH OR SALT WATER

Fresh water is a small part of the total volume of water on the planet, compared with the vast quantities of water in the oceans. Sea and fresh water present very different problems to the organisms living in them.

'Water, water, everywhere, Nor any drop to drink,' cries the Ancient Mariner, in Coleridge's epic poem, from a vast ocean where no rain or dew falls to quench his thirst. Most freshwater organisms cannot live in salt water, nor sea water organisms in fresh water. Only a few, specially adapted, can migrate from one to the other. The invisible barrier is the salt in the sea: it is a solution of the salts of sodium (Na), calcium (Ca), and magnesium (Mg), of which common salt (sodium chloride) is by far the most abundant. The concentration of salts in the water is about 35 parts per thousand, similar to the cell contents of most organisms, so that simple celled organisms living in the sea are always able to exchange water and salts through their surface membrane.

However, if one of these organisms is put into fresh water, water passes from the less concentrated solution (the freshwater medium) into the more concentrated medium (the cell contents) through the semi-permeable cell wall, until either the solutions inside and outside are equal in concentration or the cell bursts. So organisms that live in fresh water have had to adapt to prevent too much water entering their cells.

How much water?

It is hard to comprehend the vastness of the oceans on our planet. They cover about 70 per cent of the surface to an average depth of over 3,800 metres: a total of 1,368 million cubic km of water. When we compare the volume of water in the oceans with that of continental fresh water, it is even more staggering. Yet most of us are only acquainted with the edge of the sea and the organisms that live in a narrow coastal belt. The amount of water in lakes and rivers is very small, though this portion is important to us, and to most of the freshwater organisms that we shall be looking at in this book. The small volume of water being carried at any one time in the atmosphere as water vapour belies its importance because the recycling is very rapid. The greatest

Oceans
(1,350 × 10¹⁵ cubic metres)

Glaciers and polar ice
(29 × 10¹⁵ cubic metres)

Underground aquifers
(8.4 × 10¹⁵ cubic metres)

Lakes and rivers
(0.2 × 10¹⁵ cubic metres)

Atmosphere
(0.013 × 10¹⁵ cubic metres)

Biosphere
(0.0006 × 10¹⁵ cubic metres)

▲ Earth from a satellite with the vast Atlantic Ocean on the left and the 'tea cup' of the Mediterranean and shores of Europe on the right. The cloud formations indicate the directions of the north-east trade winds and the spiral cyclone shows where warm air from the south meets cold air from the polar front.

▶ An iceberg carried south from the Greenland pack ice into warmer seas begins to melt and so floats higher in the water. Two earlier float levels are indicated by the shelves on its profile, as well as the different patterns of weathering and melting in air and water.

▶ Water drops extruded from the tips of grasses in the cool night air. In the day, water is given off plants as invisible water vapour—part of the cycling of fresh water on which we all depend.

◀ Each circle represents the volume of water that is contained in each reservoir on our planet at any one time.

volume of fresh water is locked up in the ice caps, mainly at the North and South Poles. If these were to melt, the sea level would rise by 60 metres, inundating much of the present land surface and most cities. There is plenty of evidence that sea levels have altered in the geological past due to changes in the volume of water in the ice caps.

During climatically warm periods higher sea levels resulted in the raised beaches and wave-cut platforms we now see far inland; whereas during glaciations the expansion of the polar ice caps caused sea levels to drop, so typical land deposits, such as peat, occur below today's sea levels. It was during one of these periods around

available for plant and animal life. The vast, continuous oceanic area holds for would-be colonizers very different problems from the discontinuous, and often isolated, bodies of fresh water on the continents. Freshwater colonizers must have a means of dispersal and a way of maintaining their position in the general downstream movement of all continental water systems. The spin of the earth drives currents in many different directions, sometimes over considerable distances: for example, the North Atlantic Drift carries warm water and organisms from Florida well to the north. In freshwater systems all material including organic and mineral nutrients tends to be carried down into estuaries or out to sea; whereas in the sea sediment settles in deep water on the sea bed. Nutrients from water deeper than 100 metres can be brought to the surface either by upwelling currents or by the deep mixing of the water during winter storms. Light cannot penetrate far through water, and the productive zone of deep oceans is restricted to the surface 100 metres or less. However there are few freshwater systems of any great depth, so that, unless the water is cloudy with suspended detritus, light can usually reach the bottom. In running water the productive zone is mostly restricted to the bottom, where shelter and sites for anchorage are available. Each body of fresh water has a unique environmental character depending on its position in the landscape and its chemical or temperature regimes. In contrast, environmental changes in the sea occur as gradients rather than sharp boundaries, so that there are fewer distinct habitats than in fresh water.

Density, temperature and thermoclines

Fresh water is less dense than salt water. In estuaries, or where ice floes are formed around the North and South Poles, the fresh water floats above the salt water. As icebergs drift into warmer seas and climates they melt. The fresh water is then mixed with the salt by currents and wind action.

Warm water is generally less dense than cold water, except when cooled to below 4°C, when the water becomes less dense again. In spring, as the sun warms the surface of both sea and lakes, a warm layer of water forms which floats above the cool deep water. The boundary between the two layers, where the water temperature drops sharply, is called the thermocline. This stratification of the water is ecologically significant, as it initially encourages and then limits the growth of surface plankton by isolating the nutrient supply in the lower layer until storms or the cooling of the surface waters cause a mixing of the layers.

50,000 years ago that man walked across what is now the English Channel into Britain, pursuing animals that included mammoths and woolly rhinos!

Contrasting conditions

The way in which salt and fresh waters are distributed on the earth's surface greatly affects the living sites that are

PROJECTS

1 Watch the effects of fresh and salt water on an egg. Put two eggs in vinegar and leave overnight. This will remove the hard outer layer of shell. Put one egg in a jar of fresh water and the other in a concentrated salt solution. Water can pass in or out of the shell membrane. What happens after about 8 hours, and why?

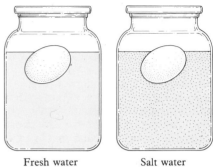

Fresh water   Salt water

2 Find a rectangular transparent container at least 25 cm long. Fill it with cold tea to around 4 cm depth. Carefully add about 2 cm of warm water, pouring it over the back of a spoon to avoid disturbing the layer of tea. Observe the thermocline between the warm and the cold water. Try tilting the container backwards and forwards. What happens? Use a hair-dryer and blow warm air across the surface. What happens to the thermocline? Finally, float a few ice cubes in the warm water and observe what happens.

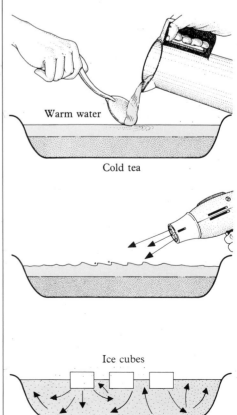

Warm water

Cold tea

Ice cubes

11

MOVING WATERS

Water is the main agent of erosion and deposition along our coasts and between highland and lowland on the land's surface, moving enormous amounts of material each year.

An experienced sailor will know the state of the distant sea by the waves that beat on the shore. A fisherman can gauge the speed, depth and bottom profile of a river by the patterns on the surface. But neither has to contend with hanging on to a wave-battered rock face or maintaining his position in a fast flowing river. Water, driven by wind or moving downhill, has potential energy. This energy may be changed into electricity in our hydroelectric stations, but if it is not harnessed it will work away at the surface it flows over, eroding, transporting, and, in more slowly moving waters, depositing the earth's substance. Friction dissipates the water's energy so that on a gently sloping shore or a wide river bed, where the water covers a large surface area, it is considerably slowed down. In a river the margins move more slowly than the centre. This effect is more easily observed on a glacier where the river of ice runs so slowly down its valley that the faster moving centre is easily distinguished from the more slowly moving, debris-strewn edges. Although you can chart the relative speeds of a river with a floating stick, you will be surprised at the result when you do the same in a rough sea. The stick will stay in almost the same place, the waves carrying it up and then passing on towards the shore, the wind acting on the surface of the sea much as it does on a field of corn.

Currents

Where winds blow in the same direction consistently (when they are known as prevailing winds), large bodies of water may be moved in the same direction, forming a current. Currents are also caused by the spin of the earth, and move in a clockwise direction in the Northern Hemisphere and in an anti-clockwise one in the Southern Hemisphere. The pattern of currents is also modified by differences in temperature and salinity and by the shape of the sea bed: in some areas the nutrients are brought to the surface, and in others they are carried into other regions.

▲Two valley glaciers 'pour' down from the ice cap Eyjafjallajökull in Iceland to meet the braided streams in the foreground carrying and depositing the material from other glacier snouts higher up the valley.

▶A storm beach of shingle, thrown up by the sea against a low-lying river valley in south-west England, contrasts with the headland of hard rock beyond.

▼A waterfall in Iceland caused by a bed of hard basalt lava that is eroded more slowly than the material above and below it.

▲Lynmouth on the day after the disaster in 1952, showing the devastation and the river's ammunition of boulders strewn around.

▶The East Lyn river in Summer 1981 showing the steep-sided wooded valleys behind and the high flood walls rebuilt since the 1952 disaster.

▼The river Cuckmere in the south of England meanders through its own alluvial deposits.

Tides

Day in, day out, the sea level rises and falls with the tides, the widest varying tides occurring at the spring and autumn equinoxes. In the oceans the tides follow a simple pattern, but where they meet the shore there are many modifications according to the type of shore. Where the high tide travels up an estuary it may hold up fresh water running down and so cause extra high tides; thus in the Severn estuary in the south-west of England the tidal range is 14 metres compared to a range of 6 to 7 metres on the open coast at Dover in the south-east. During the equinoctial tides the water piles up to form a gigantic wave known as the Severn Bore. A similar but much more devastating wave may be caused by unusual weather conditions coinciding with the spring tides in the North Sea. In small basins, such as the Mediterranean, the tide varies little.

Water at work

Demolition The river Lyn threads its way through picturesque wooded valleys on the Exmoor plateau, in south-west England, to emerge at Lynmouth. It collects its waters from the open moorlands of Exmoor, formed on tough, resistant rock that is not permeable to water, and Exmoor receives most of the rain that the south-westerly winds carry off the Atlantic. On the night of 15 August 1952 a freak storm released between 23 and 28 cm of

rain on the moor. The small streams became raging torrents, carrying soil, plants and boulders into the valleys, where they met other tributaries carrying similar loads. The steep valley falls 50 metres in 6.5 km, and the water, loaded with trees, debris and vast boulders up to 10 tonnes, destroyed everything in its path. The Lynmouth disaster was a tragedy, but the steep, V-shaped valleys of the highland regions of Europe were all carved out in this way before history could tell the tale. During any year 10 to 50 tonnes of material per square km are removed from the land surface and deposited in lowland regions or the sea.

There are many areas along our coasts where the sea eats at the base of cliffs, which crumble and slip from above, so losing ground annually. Where the rocks are tough, resistant material, such as granite, erosion may take place very slowly, but where the material is soft, such as chalk, a metre a year can be lost. The shape of the coastline is governed to a large extent by the sea acting on harder or softer rocks or on areas of weakness in the faults and folds of the earth's crust. Where the waves have cut a platform on bedrock, uneven erosion along planes of weakness in the rock, or around boulders, gives rise to that particularly fascinating sea-shore habitat, the rock pool. Water may dissolve the bedrock it passes over, particularly rocks with chalk or limestone in them, sometimes forming underground water courses, caverns and blow-holes. One of the major agents of water demolition that shaped our landscape was ice, both through frost shattering of surface material and through the carving action of glaciers in their U-shaped valleys.

Deposition Much of our present landmass is composed of rocks that were deposited from water or ice as sediment in past geological ages. The material that was carved away by the glaciers was dumped as moraine at the snout of the glacier, or in vast outwash plains, which today are farmland with relict water bodies like the meres. This process is still going on wherever water is at work demolishing old habitats and creating new ones. As soon as the velocity of a body of water falls, it drops material. Velocity is most frequently associated with gradient, and wherever this becomes more gentle, as on a sloping shore or plain, the deposits accumulate. First the larger material is dropped, then progressively smaller particles as the velocity decreases, until only the very fine particles of silt are left in suspension. These drop out in the almost still water of lakes or estuaries.

On the shore the pattern of deposition is controlled by the slope of the shore and the prevailing wind and wave direction. The resulting longshore drift of material is graded from boulders and shingle on steeper shores to sand on gently sloping beaches. Sudden changes in the force and direction of winds may entirely alter the shoreline and deposits.

In streams and rivers the pattern of deposition is very different, as all material dropped obstructs the path of the river, so that the river is constantly cutting a route through its own sediment. But the route is not a straight one. A stream or river tends to meander along gentle sloping valleys in sinuous curves where erosion takes place on the outside edge of the curve and deposition on the inside edge; sometimes the curves increase and expand until the water cuts through, leaving an oxbow lake, and the whole process begins again.

PROJECTS

① Shake a jar half filled with sand with a strong solution of potassium permanganate or with dental indicator and leave overnight. Find a place where a freshwater stream cuts across a sandy beach and place spoonfuls of the coloured sand in different situations of the stream; centre, banks, shoaling areas, curve, etc. What happens to the coloured grains?

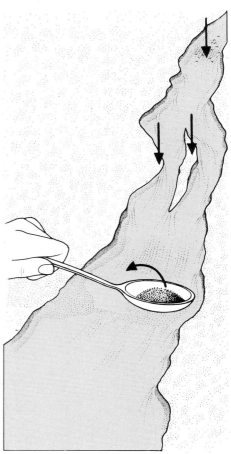

② After a heavy rainfall visit your local stream. Is the water cloudy or clear? What types of sediment are apparent? If the water is cloudy collect some in a jar or bottle and watch it settle. Is it similar to any of the deposits occurring in your stream? In what order of size does it settle?

EQUIPMENT TO TAKE OR MAKE

Some of the equipment that is frequently used for environmental recording is described here; other equipment is mentioned where it is used for particular projects.

Most of the organisms mentioned in this book can be observed with the naked eye once they are isolated in a glass jar. But finding them in their environment often requires experience and visual aids, as our eyes are not adjusted to seeing below the surface of the water. Once we can see, we need equipment to collect and observe the animals without damaging them, and books to help to identify them. Now we can begin to ask the interesting questions, such as 'what governs the distribution of any animal or plant?' 'Is it a predator/prey relationship or an environmental factor?' To answer the questions we need equipment with which to investigate and record the environmental factors.

Seeing

There are numerous types of lenses to make the distant nearer or to enlarge the minute. BINOCULARS come in all shapes, sizes and weights, but it is most important to select a pair that is easy to use because of the small field of view. 8 (magnification) × 40 mm (lens diameter) is a useful combination for naturalists. A HAND LENS is an essential piece of equipment for looking at the minute in plant or animal material. A magnification of × 10 is easy to use; higher magnifications require practice. For looking at small mobile animals it is possible to buy a useful boxed lens called a magnispector. With this you can watch the behaviour of an animal without damaging it or critically altering its temperature (you are warm blooded and have a much higher temperature than it has). For watching animals in their habitat we suggest that you make an UNDERWATER VIEWER, as shown on p. 150. This gets rid of surface reflections in fresh or salt water environments. MICROSCOPES are very expensive aids, but reveal another world of structure and adaptation. They also require other specialized equipment which is not in the scope of this book. However, Osmiroid make a simple swivel-lens tripod giving a × 20 magnification, which is very useful.

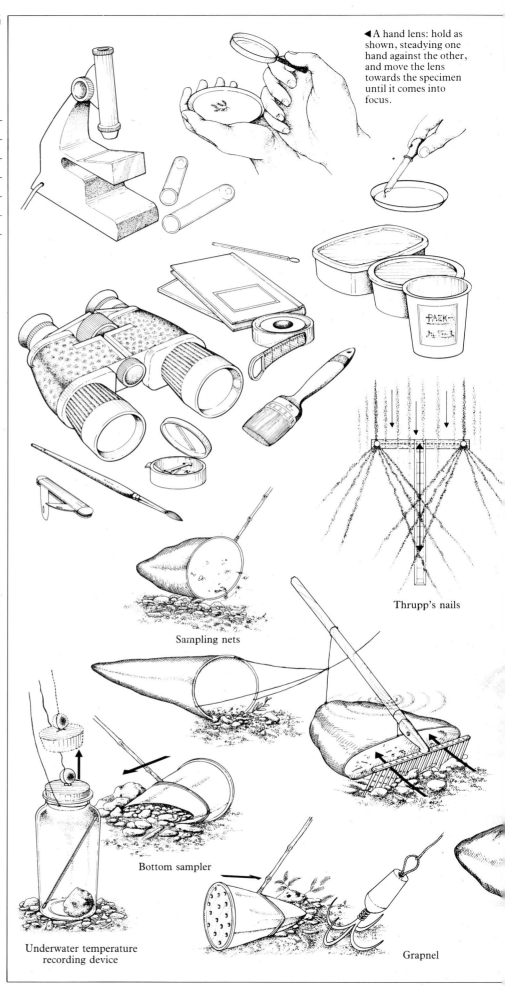

◀ A hand lens: hold as shown, steadying one hand against the other, and move the lens towards the specimen until it comes into focus.

Thrupp's nails

Sampling nets

Bottom sampler

Underwater temperature recording device

Grapnel

14

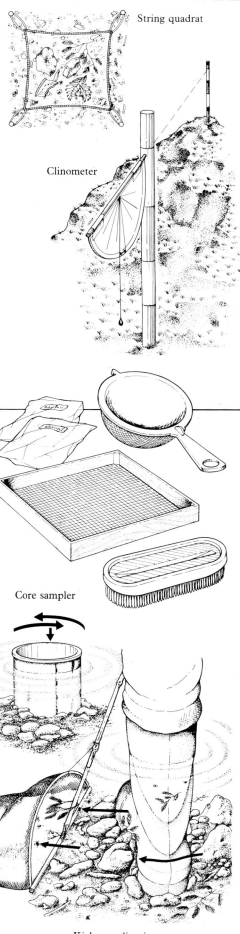

String quadrat

Clinometer

Core sampler

Kick sampling in a stream

Collecting

Much useful collecting and sorting equipment can be obtained by saving food containers, or adapting kitchen equipment when you need it. NETS are essential for fresh and salt water organisms. You can buy a sampling net, but you can make your own with some stiff wire, a broom-handle and some nylon curtain material with a mesh of around 1 mm. You can then change the shape of the net according to whether you are sampling at the bottom of the water or in the centre. Sampling on the bottom of fast streams requires a special technique known as 'kick sampling': you place the net downstream of where you kick so that you catch the dislodged animals in your net. For pond and sea water environments a shrimping net is useful for sampling the bottom. Alternatively you can attach a bag to a garden rake and gently disturb the bottom so that you can scoop up any of the organisms living there. In the water itself move the net slowly in a figure of eight across an area of water so that the water passes through the net but the animals are trapped in it. TINS large and small can be adapted as core samplers—both ends removed from a large tin—or as a sampling device—holes punched in the bottom of a small tin, end flattened slightly, and pole attached. SIEVES of all kinds are useful to get rid of mud and fine detritus as well as for catching your animals in your aquarium. Once you have a sample, put it on a white enamel or plastic TRAY to distinguish the smaller moving organisms against the white background. PIPETTES are useful for catching the organisms, and make sure you have numerous lidded glass or plastic POTS into which to sort them. It is particularly important to separate carnivores (usually fast-moving, long-legged, big-eyed) from herbivores (usually slow-moving, legs short or absent, eyes reduced), or you might find your sample considerably depleted. Mark each pot and your notebook with a specimen number. Nothing is more annoying than losing track of a sample. To obtain plants in deep water a GRAPNEL is useful. This can be made from four pieces of thick wire with a lead or other weight attached and plenty of smooth string that does not tangle easily. Plants are better kept in individual POLYTHENE BAGS with a LABEL. Other useful articles are a SCRUBBING-BRUSH for stone sampling (p. 73), PAINTBRUSHES, a PENKNIFE and a variety of SPECIMEN TUBES.

Recording

First, a good MAP: buy the largest scale you can, and ensure that rivers and still waters are marked. A hard-backed NOTEBOOK is essential for serious observers, so that you can check up on earlier observations. TAPE MEASURES of 20 to 100 metres are useful for transects, but STRING marked into metre sections is as good. You will need a COMPASS for any transects. QUADRATS can be made in wood or string in various sizes according to the type of habitat that you are investigating. If you need to measure the angle of slope you need a CLINOMETER. There are gun clinometers on the market, but you are only measuring the angle from the horizontal, and can make your own with a straw, a protractor with a hole drilled at the 90° intersection, and a weight on a string. Temperature is easily recorded with a THERMOMETER at the surface but is more difficult to record in water at different depths. However if you find a bottle with a glass (or easily removed) stopper that you can fit your thermometer in, you can make a device that you can lower to different depths in the water, pull the stopper out so that it fills with water, pull the apparatus up and record the temperature. There is on the market a variety of tough environmental recording thermometers. Speed of water at the surface can be measured comparatively with a device known as THRUPP'S NAILS. Make a T-shape in wood and attach a centimetre scale on the upright. Fix 2 5cm nails 10.2cm apart into the cross piece, leaving them to protrude at least 3 cm above the wood. Place this in the current with the scale downstream of the nails. Keeping the wood below the surface, read the lowest point at which the current waves from the nails overlap on the scale. What happens in the fastest currents?

Waterproof light meters are very expensive, but try using a SECCHI DISC. This was invented by an Italian admiral who lowered a dinner plate over the side of the boat until it disappeared from view, took a reading, raised it until it reappeared and took the mean of the two readings to obtain the visibility; use a white disc, 25 cm in diameter, and a marked string. Finally, pH—the most difficult of all factors to measure without special equipment, but pH papers with different ranges are available, as well as a kit that lasts longer and is more reliable. A few essential nutrients such as nitrogen, phosphorus and potassium can be tested with special kits.

Identifying

On p.188 we suggest some useful books, and keys related to specific habitats.

Please remember, when you have collected all your data, to return all organisms to their environment. However small, they are an essential part of a living community.

STILL WATERS

Each pond or lake has a characteristic community of plants and animals, as well as its own particular set of conditions which are determined by landscape and climate.

The rustling of the reeds around a pond, or the plop of a fish breaking the surface to catch an aerial insect, only emphasizes the mysterious quiet of a pool's surface that contrasts with the clamorous noise of a rushing stream. Yet most still waters are dependent on running water to maintain them, and an inflow and outflow can be observed. The time taken for water equivalent to the volume in the lake to pass from inflow to outflow is known as the replacement time. In highland areas, ponds and lakes usually have a shorter replacement time than in lowland areas, but they also fill up with sediment from the fast-flowing streams at a greater rate. All ponds and lakes have a tendency to get filled in (pp. 48–9). In lowland areas of Europe many open water bodies have disappeared over the last three centuries, due to natural infilling and to increased drainage of the surrounding land.

How was it formed?

Each still water body is dependent for its existence on a combination of climate, topography and geology. In hot climates evaporation exceeds precipitation for most of the year so that most smaller freshwater bodies are temporary features of the landscape. The rate of evaporation depends on the surface area of the lake, so deep lakes with a small surface area are always a permanent feature. In temperate climates rainfall is high for much of the year and there are many more permanent water bodies. One of the factors that has influenced the distribution of water bodies in Northern Europe is the action of the ice sheets in the last 600,000 years which scoured out large valleys and dumped mounds of moraine in front of the ice. Because sea levels were lower during the Ice Ages many of these valleys are lower than today's sea level. The retreating ice also covered large parts of lowland Britain and Northern Europe with extensive sheets of till, or boulder clay, which is often impermeable to water. These areas may hold reservoirs of ground water that appear above the

▶Mývatn, a lake in central Iceland, was formed around 6,000 years ago when lava dammed up the river Laxá. Around 2,000 years ago, lava flowing out of the volcano (in the top left of the picture) formed these pseudo-craters on entering the marshy ground around the lake, creating a more extensive lake dotted with islands that now support many thousands of breeding birds during the short Icelandic summer.

▲Crosemere is a lowland lake in north-west England surrounded by pasture and arable farmland, which give it a eutrophic character. This winter picture shows the sedge and alder carr in the foreground but the reed fringe around the lake is obvious only in summer.

▶Llyn y Fan Fach is an oligotrophic lake in the mountains of south Wales. It lies in a corrie that was carved out from the hill by the action of ice. The lake is fed by nutrient-poor water running off the hills around.

▶ The maximum extent of glaciation during the last 600,000 years in Europe and America

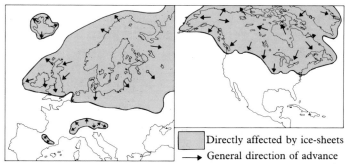

Directly affected by ice-sheets
→ General direction of advance

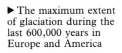
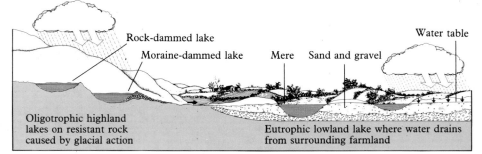

Rock-dammed lake
Moraine-dammed lake
Mere
Sand and gravel
Water table

Oligotrophic highland lakes on resistant rock caused by glacial action

Eutrophic lowland lake where water drains from surrounding farmland

surface as meres, but lack any surface inflow or outflow channels. Those that are very deep are thought to have been formed from isolated blocks of ice that remained in the till, and on melting left kettle holes. The geology makes a considerable difference to the nature of water bodies, both in the soluble minerals, such as lime, that a rock may contain, and in the hardness or permeability of a rock which influences the type of landscape and the sediment on the bottom of the pool or lake. Changes in the earth's crust may also cause lakes to appear where faulting produces lines of weakness, along which valleys have formed, such as the upper Rhine valley in Germany or Loch Ness in the Highland region of Scotland. Much more recently flows of lava in Iceland around 2,000 years ago dammed up the lake at Mývatn, one of the richest sites for freshwater birds in Europe.

How big is it?

The size of a permanent still water body may vary from a well or water-butt to a lake many hundreds of kilometres long, such as Lake Nyasa. The size and depth of water will influence the species that live there as well as the temperature of the water and its seasonal fluctuations. In ponds and small pools the temperature will fluctuate fairly rapidly with the air temperature, although once a film of ice covers the surface further lowering of the temperature continues more slowly. In deep lakes another factor operates; the surface layers are warmed in the summer but the colder, denser water remains below, a sharp thermocline separating them. Oxygen is also in short supply where there are no green plants to aerate the water. In addition all organic nutrient material accumulates on the bottom in the cold layer and is unavailable to organisms in the upper layers. With a drop in temperature in the autumn the upper layers of the lake cool, and once the temperature falls below that of the

lower layer there is a massive turnover of water and nutrients in the lake, of which plant and animal life can take advantage as soon as the warm spring days begin. The depth of a water body and its marginal shape also influence the plants that can grow there, so that shallow lakes can support extensive reedswamp, whereas deeper lakes can only support a floating and/or a submerged zone, the latter being dependent on clear water to allow the light to penetrate.

What grows in it?

The character of each lake or pond is brought about by the relationship between the animals and plants living in it and the nutrient and mineral supply that is available from the surrounding catchment area. Within different regions of rock type and climate we can detect some typical lakes, and these are often characterized by individual associations of animals and plants. In the highland regions, where lakes are formed on hard resistant rocks that give few nutrients when they are eroded, the lake water is clear and the bottom covered with gravel or pebbles, yet there are few plants in evidence and even the reed fringe is sparse. These lakes are known as oligotrophic, or 'few-feeding'. In contrast, lowland lakes are usually eutrophic, or 'well-feeding', when the inflow water has passed through much agricultural land and is laden with nutrients. If there is lime present the pH of the water is increased, making it a richer place for plant and animal life. These lakes are characterized by a thick fringe of reeds, a floating mat of waterlily and pondweeds and an abundance of free-floating algae and duckweeds. There are many types of lake, characteristic of particular conditions (some of these are described on pp. 48–9); but the remarkable thing about each one investigated is its unique character and its capacity to change from season to season and from year to year.

◄This Icelandic ice lake is formed at the snout of a retreating glacier which leaves morainic mounds that dam up the melting ice. At the edge of the glacier the ice may break off in great slabs that melt slowly in the lake.

PROJECTS

1 Make your own kettle holes. Place some ice cubes, well spaced out, on the bottom of a pie dish. Fill it with enough coarse sand to cover the ice. Stand the dish on a level surface. As the ice melts, crater-like hollows will form in the cubes, with a mini-mere at the bottom.

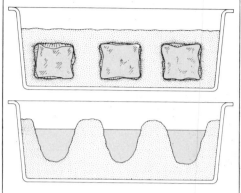

2 Use ice as an insulating layer. Find a heat-resistant glass or enamel dish, about 10 cm deep, and a cardboard box that it will fit into easily. Fill the cardboard box with enough crumpled or shredded newspaper to surround the surfaces of the dish completely, in order to insulate them from heat loss, and place the dish in it. Fill the dish with cold tea to 8 cm in depth, and put the whole thing in the freezer. Measure the temperature at the surface, then at 4 cm and 8 cm below the surface, every 15 minutes. What happens before and after ice forms on the surface? Take the box out of the freezer and remove the dish so that you can watch what happens when you top up the dish with cold, clear water.

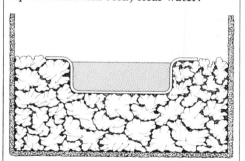

3 If you have safe access to a lake and a boat, try recording the temperatures at different depths in the lake and at different seasons using the thermometer-in-bottle technique described on page 15. Make sure that the string is marked in centimetres and metres so that you can determine the depth at which you are recording each temperature. Can you find the thermocline in summer? What is the temperature at the bottom, and on the surface of the lake? How do your results compare with the temperatures in a small pond?

PONDS AND POOLS

A lowland pool such as this one at Wicken Fen conceals an action-packed space below its quiet, reflective surface, for there is abundant food in this eutrophic, or 'well-feeding', habitat. The emergent and floating plants give us a clue to this condition, for the tall stems of the emergent reeds only grow in nutrient-rich conditions and will produce much more green food in a single season than any grain crop. The floating pads of waterlily leaves obscure the lettuce-like submerged leaves and thick stems that are rooted in the mud, all these being the homes and hunting places of multitudes of freshwater invertebrates. Finally the lush fringe of willow carr presses in around the pool, needing only a tussock of grass or sedge in which to anchor, yet each year it adds to the organic nutrients in the pond at leaf-fall. But willow shrubs are eaten by many different invertebrate species as well as by grazing animals. Many lowland pools are grazed and farmed to their edges so that typical willow or alder carr never develops.

Wicken Fen in Cambridgeshire in the east of England with a fringe of common reed *Phragmites australis* and yellow waterlily *Nuphar lutea* and willow carr beyond.

INVADERS AND OPPORTUNISTS

Even before man began to have such a dramatic impact on freshwater environments, many aquatic plants and animals depended on efficient methods of dispersal and the ability to exploit new habitats as soon as they became available.

Dig a pond, dam a valley or even put a bowl of water on a window-ledge, and within a few days living organisms will have taken up residence. Algae, either free-floating or in the form of scum on the surface, start growing; protozoans appear; and later the immigration of larger organisms, such as rotifers, water fleas and water boatmen, occurs. Like weeds which appear on newly dug soil, there are aquatic animals and plants which show a remarkable facility for travel and an equally remarkable ability to grow profusely and then move on again when other, more competitive species, which tend to have less efficient dispersal mechanisms, move in.

Aerial dispersal

Movement from one body of water to another requires an ability to traverse tracts of dry land which are totally inimical to the survival of an animal or plant. Travel may be by air, either by blowing along as a fine dust of drought-resistant spores or seeds, or by actual flight, as in the adult stages of many aquatic insects. The tiny offspring of the water spider climb to the top of the surrounding plants and weave gossamer threads which the wind whirls into the air, together with the spiders, and scatters all across the countryside. Thousands probably die for every one that, by chance, reaches a new pond.

Occasionally freak conditions transport the most unexpected animals: rains of frogs and even fish have been recorded. Normally, however, aerial dispersion relies on stages that are very tiny and very light, and so minute are the chances of even one spore or one tiny reproductive cyst reaching a new body of water that they have to be produced by the million. The few successful colonizers then have to reproduce very rapidly, often by asexual budding or division, so that in the limited time available enough of the dispersal stage can be produced.

◄One method of transport for the floating duckweed *Lemna minor* is on the back of a toad *Bufo bufo* which is hopping across a canal towpath.

►The thread-like filaments of the alga *Spirogyra* are frequent colonizers of new water bodies. The filament on the left is in a growing state with the cell contents filling the cell and the flat, ribbon-like chloroplast spiralled around the cell wall. The two filaments on the right are conjugating: the cells of each filament are joined by a tube and the cell contents condense and pass through the tube to fuse with the cell contents of the neighbouring filament. They form a hard-walled resistant zygote, or egg cell, that can resist drought and be blown or carried to new places.

◄The seed heads of the unbranched bur reed *Sparganium emersum* are borne on aerial stems that emerge above the floating, strap-shaped leaves. The buoyant corky fruits float on the water for weeks and are carried to new places in the current or are eaten and pass through the digestive system of water birds to be deposited in another watery habitat.

◄Reedmace *Typha angustifolia* is an emergent water plant that invades the margins of still or slow-flowing waters. Each flowering spike produces many hundreds of tiny plumed seeds that are dispersed by the wind.

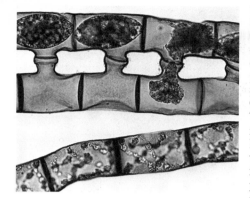

Hitching a lift

Mud cleaned off a pair of boots after wading through a pond will grow a number of weeds, and if it is immersed in water a variety of animals will also hatch out. Under normal conditions, all *Daphnia* (water fleas) are female and reproduce by parthenogenesis (that is, by unfertilized eggs), but as conditions deteriorate, either through over-crowding or through desiccation of the pond, some of the females produce male offspring. These then fertilize other females which in turn produce ephippia, or resting eggs. These eggs can still hatch after a resting period of several weeks or months, and can even survive drying. Some of the inhabitants of temporary pools, like the fairy shrimp *Chirocephalus* (pp. 36–7), produce resting eggs which actually require a drying-out period before they will hatch again on re-wetting. They may travel in mud caking the feet of water birds, or these birds may even collect the eggs on their feathers during a dust bath, and carry them that way.

Many mud-dwelling plants have small, mucilaginous seeds which get very sticky when moistened and attach themselves to a passing object; several rushes are dispersed in this way. Other plants have seeds which can pass right through the digestive tracts of birds or mammals and still remain viable; a visit to a sewage farm will soon prove the extraordinary ability of the tomato to travel in this way.

Water dispersal

The seeds of many aquatic plants, not unexpectedly, are dispersed by water. Both water plantain and arrowhead have seeds that are fitted out with floats filled with air. The test of the yellow flag has a thin outer layer. Once this layer rots the seed sinks and germination begins.

It is not always the seed that is the dispersal stage. Twigs of willows broken off by flood water can root and grow in the mud of a river bank. At times of flood, whole plants of duckweed and water fern, which normally inhabit still backwaters, may get scattered widely and be left behind in new water bodies by the receding

waters. They can also be carried on the backs of amphibians moving from one water body to another. Vegetative reproduction can be a powerful method of dispersal in waterways when a new plant is formed from a fragment of the parent plant. The rapid spread of the Canadian pondweed is a good example. This plant is dioecious (having both male and female plants) in its native North America, but only female plants were introduced in Ireland in 1836, from where they were taken to many parts of Europe during the nineteenth century. It spread rapidly through the canal system, any fragment that broke off soon rooting and growing into a new plant.

Animals which are totally aquatic are much slower to spread. Pond snails such as the wandering snail are able to spread, transported by birds, to new waters either as spawn or as newly hatched little snails. Consequently, they are almost cosmopolitan in their zoogeographic range. When there is rain or heavy dew to keep the vegetation wet, fish such as the eel are able to crawl overland to new environments. Amphibians (newts, frogs and toads) simply walk over land (pp. 26–7). Fish species tend to be the same throughout a river system, spreading by their own locomotory powers, or during floods. The pattern of distribution of char in isolated, deep, cold-water lakes in many European countries today shows which of these lakes were interlinked during the last glaciation, some 12,000–14,000 years ago. The isolation of these lakes has resulted in the development of various forms of the char, each with a distinct local name. The herbivorous feeding habits of many coarse fish allow them to move in rapidly and exploit new habitats; great opportunists, for example the roach, can undergo huge population outbursts when conditions are right. The goldfish is an opportunist *par excellence* in being able to inhabit and exploit polluted waters which are deficient in dissolved oxygen. Generally speaking, herbivores are limited not by food but by factors such as overcrowding or predation. Predators are generally limited by the availability of food, hence they tend to be less effective than herbivores as exploiters.

Dispersal in action

Perhaps the most convincing testament to the ability of aquatic animals to travel was the observation of Professor Thienemann in North Germany. He recorded the animals occurring in a tank of water over four successive summers, starting afresh each spring. In total 103 species occurred, only 10 of which were recorded in all four years. Most remarkable were the 8 species that were totally aquatic and possessed no obvious method of dispersal.

1 Use a species of common pond snail to investigate the effects of food supply and space on snail size and number of eggs laid. Set up two 500gm jars, and two 1kg jars, all containing pond water. Place a piece of Canadian pondweed in one jar of each size, and twice as much in the other two. Keep these on a sunny window-sill and make sure the weed is growing well before putting two snails in each jar. Snails are hermaphrodite, and will fertilize each other. The eggs are laid in a jelly-like mass, in which you can count the number of embryos. The increase in size of your snails can be assessed by weight, or by measurement of the diameter or length of their shells, depending on the species that you have chosen for the experiment. What conclusions do you come to concerning the space and food needs of your snails?

The great pond snail *Lymnaea stagnalis* with the eggs of another snail across its shell.

2 Repeat Thienemann's experiments, placing a 50-litre (or larger) tank outside and regularly recording its inhabitants with the aid of a microscope or lens for the smaller organisms. Lay out your record card so that you can see easily which are new occupants and which disappear. Can you find the nearest freshwater pond, pool or lake from which the invaders may have come? Record only the different groups of organisms, as many of the species or genera are difficult to identify without expert help.

OXYGEN FOR LIFE

Aquatic animals, like all other animals, require oxygen. Ensuring a continual supply of oxygen is a major problem when living under water.

There are two ways of resolving the problem. One is to take a supply of air from the surface down under the water. The other is to have a thin body wall, or a specialized region which can extract oxygen from the surrounding water. Many larvae have evolved thin-walled gills for gaseous exchange.

Surfacing for air

Animals which have to return to the water surface to draw in a fresh air supply can be considered as being partially aquatic. Both the larvae and the adults of the diving beetle are lighter than water and naturally bob up to the surface like a cork, unless they keep swimming or cling on to underwater plants. The adult beetles rise to the surface tail first, and trap a bubble of air beneath their wing cases, or elytra. The silver water beetle uses a different system: the underside of its body is covered with tiny water-repellent (hydrofuge) hairs which trap a layer of air, giving it a silvery appearance. In contrast to the diving beetles, it rises to the surface head first. It breaks the surface film with its antennae, which are also covered with hydrofuge hairs and form a connection between the atmosphere and the layer of air on its underside. This air layer also functions like a gill, drawing in more oxygen from the surrounding water as the beetle uses it up.

Water boatmen, or backswimmers, are also buoyant because they carry an air supply trapped in hydrofuge hairs on the underside of the body. A backswimmer rising to the surface breaks the film with the tip of its abdomen, which is surrounded by a fringe of long hairs. This connects the atmosphere to the paired openings — the spiracles — which open into the internal branching system of tubes or tracheae within the backswimmer's body.

The great pond snail can often be seen crawling upside down on the surface film, taking air into its mantle chamber within the shell. This chamber is lined with copious blood vessels and functions like a lung. When the pond is iced over the snails can no longer reach

▶The water scorpion *Nepa cinerea* takes in the air through its terminal breathing tube, whereas the red parasitic larval water mites, attached to its body, tap oxygen dissolved in the water through their thin-walled bodies.

▲A great pond snail *Lymnaea stagnalis* rises to the surface to take in air.

▶The silver coating surrounding the body of the water spider *Argyroneta aquatica* is a layer of air which it carries from the surface down to its underwater airbell.

◀This mayfly nymph *Ecdyonurus torrentis* lives in well oxygenated streams, extracting oxygen from the water via the paired leaf-like gills. Compare the shape (and markings) of this flattened nymph with another species, illustrated on p. 70, collected from the same site.

the air, so they move down to the bottom and rely on what oxygen they can extract from the water through the skin covering the extended foot and head.

In both the water scorpion and the water stick insect, the end of the abdomen is extended into a snorkel-like breathing tube which is really two half-tubes held together with interlocking

bristles.

The larva of the drone fly, the rat-tailed maggot, also has a snorkel-like breathing tube on the end of its abdomen, but this is very extensible. The tip of this telescopic breathing siphon is kept above the water, and its fringe of eight feathery hydrofuge hairs prevents water from entering the siphon. The fully extended siphon is

▲Larva (left) and pupa (right) of mosquito *Culex pipiens* hanging from the water surface where they take in atmospheric oxygen through their respiratory siphons.

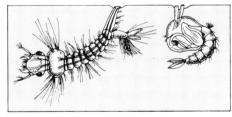

| Hairs separated by surface tension pores, spiracle exposed |
| Hairs close over spiracle, preventing entry of water |

◀Detail of hydrofuge hairs surrounding a spiracle or breathing pore at the tip of a respiratory siphon, showing how the passive movement of the hairs opens and closes the spiracle opening.

several times longer than the dirty white body, and the larva is able to inhabit the most grossly polluted water devoid of oxygen.

Both larval and pupal stages of the mosquito hang down from the water surface taking in air through small breathing tubes. One way of controlling mosquitoes is simply to pour a layer of oil over the water. This lowers the surface tension, allowing oil to enter and block up the breathing pores, so suffocating the larvae.

Air on tap

By carrying an aqua-lung filled with compressed air, man can extend the length of time he spends under water before he needs to surface. The water spider also increases its submergence time by building a platform of silk between water plants. Rising to the surface tail-end first, the spider then traps an air bubble on its underside, swims down beneath the silk web and releases the bubble, which makes the silk web bulge upwards to form a bell. The water spider has to make many repeat journeys before the bell is filled with air. It can then sit in the bell throughout the day, moving only to pounce on a hapless insect, whereas at night it goes off on feeding excursions.

Air from the water

Thin-walled animals, such as worms and small larvae, do not need to migrate up to the surface to replenish their air supplies, since oxygen diffuses through their skin. Water fleas beat their limbs continuously, thereby constantly bringing a fresh supply of oxygenated water across their body surface.

Many insect larvae or nymphs possess gills which increase the surface area for gaseous exchange. Damselfly nymphs have three flattened, leaf-like gills projecting from the end of their abdomen, while mayfly and alderfly nymphs have a series of paired abdominal gills. If mayfly nymphs encounter water with a low oxygen content, they beat their gills more quickly so as to create a stronger current passing over them. Nymphs of large hawker dragonflies have gills within the rectum and so they continually pump water in and out of the anus.

Organic material accumulating on a muddy bottom is broken down by microbes which use up the available oxygen, leaving a black anaerobic layer smelling of hydrogen sulphide. This anoxic mud is completely inimical to most animals. However, *Tubifex* worms and blood worms — the red larvae of chironomid flies — possess the red blood pigment haemoglobin, which has great affinity for the little oxygen remaining, forming oxy-haemogloblin which is then carried in the haemolymph (or blood) around the body. *Tubifex* worms live head down in tubes in bottom mud. They extend their tails up into the water and, by beating them sinuously, draw a current of water down across the body. If the oxygen content of the water is lowered, the worms increase the area of their body surface for absorption by simply backing further out of their tubes.

PROJECTS

① Fill a jam jar two-thirds full of pond water. Collect a water boatman and some pondweeds, and transfer them to the jar. Time, over a half-hour period, how long the water boatman spends under water before it surfaces to breathe. Repeat the experiment with boiled water which has been cooled to the ambient temperature. Compare the frequencies of ascent.

A water boatman *Notonecta glauca* hanging down from the water surface, taking in air, with its reflection above. Notice the length of the hind legs, which are used for propulsion through the water.

② Visit a pond or a canal and watch the way in which great pond snails take in air at the surface. How long do they spend moving around on the underside of the surface film?

③ Find a rat-tailed maggot and put it in a tall glass jar containing 5 cm of clear water. Continue adding water so that the depth increases by 2cm increments. Watch to see how the telescopic siphon increases its length with the depth of water.

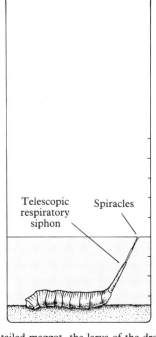

Telescopic respiratory siphon | Spiracles

A rat-tailed maggot, the larva of the drone fly *Eristalis tenax*, breathes air by extending its respiratory siphon up to the surface.

MOVING AROUND

Pond life is by no means restricted to swimming under water; some animals live on the surface film, while others emerge to lead an aerial life.

Animals which are able to move do so to find food, a mate, or an egg-laying site; to escape from predators; or to disperse. The speed of movement, however, will vary from animal to animal depending on its mode of life—herbivores tend to browse leisurely once on their food plants, whereas even the most stealthy carnivore tends to attack suddenly.

Underwater propulsion

The way in which aquatic animals move under water is linked to the way in which they breathe (pp. 22–3). Animals which extract oxygen directly from the water are denser than water and move around by crawling over the bottom or on submerged plants. Freshwater flatworms, although small in size, are able to move relatively rapidly by means of the mass of tiny hair-like cilia on the underside of the body. Leeches usually use their suckers at each end to loop along, but they can also, when necessary, swim rapidly through mid-water by snake-like, vertical undulations of their bodies. Various fly larvae move by flexing their whole bodies from side to side, which results in a much slower movement.

Like terrestrial insects, caddis larvae crawl along by moving three of their six legs at a time, two on one side (front and back) and one on the other side (middle). Large dragonfly nymphs crawl along using their legs, but if they want to move fast, either to escape or to chase their prey, they can jet-propel themselves through the water by drawing in and then forcibly ejecting water at their tail ends.

Many larvae and adults of freshwater bugs and beetles swim freely, using their hind legs as paddles. Not only are these legs flattened, but long, hairy fringes splay out to increase the size of the paddles, and hence the power of the propulsive stroke; they flatten back ('feather') on the recovery stroke. Diving beetles and the backswimmer move their hind pair of legs backwards in unison like an oarsman sculling, whereas the silver water beetle moves the middle and hind pairs of legs out of phase with each other, so that at any

▲ The medicinal leech *Hirudo medicinalis* uses its small anterior and large posterior suckers for looping over a substrate.

▼ The great diving beetle *Dytiscus marginalis* using its powerful pair of hind legs to propel itself through the water.

▶ The medicinal leech *Hirudo medicinalis* swims through the water by elongating its body and moving it in snake-like undulations.

▲ Springtails *Podura aquatica* cluster together on the water surface.

▶ Sunlight glints from the dimples made by a pond skater *Gerris gibbifer* resting on the water surface.

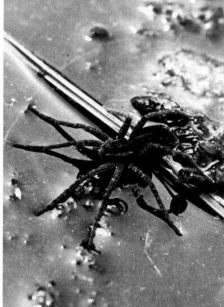

◀ An adult mayfly or spinner *Ephemera danica* rests on an oak leaf.

▲ A swamp spider *Dolomedes fimbriatus* rests on the surface of a boggy pool.

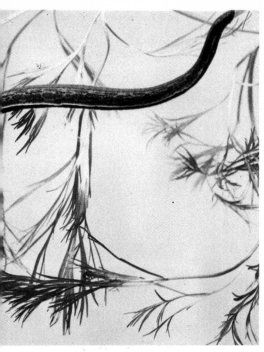

one time one leg moves forward and one backwards on each side of the body.

A water boatman, surfacing to breathe, rests head down at an angle of 30° to the surface, but, as it thrusts backwards with its hind legs to dive, the head moves downwards so the angle increases to 50°. Because it is so buoyant, it has to make a series of rapid swimming strokes to stop itself from bobbing back to the surface again.

The speed of movement under water depends on how rapidly successive strokes are made; the water beetle *Acilius* can swim at a speed of 35 cm per second.

Walking on water

Some animals spend their lives moving on the surface film. When seen with the naked eye, tiny springtails resemble specks of soot, but they often occur in dense concentrations. They have a water-repellent cuticle which keeps them dry and they are anchored to the water surface by the wettable first abdominal segment. If danger threatens, their hinged tails enable

them to jump as much as 30 cm.

Pond skaters are the water walkers *par excellence*. Their narrow bodies are prevented from getting wet by an upper waxy coating. Underneath, a dense pile of hydrofuge hairs traps a layer of air. The body is also supported by three pairs of legs; the front pair is short and the other two are long. The tip of each leg creates a dimple in the water surface. The surface tension produced by the strong attraction between the molecules of the water keeps the skaters afloat. Pond skaters row themselves over the surface using their middle pair of legs.

Other insects which walk on water include the water cricket and the water measurer.

Whirligig beetles paddle along with their bodies half in and half out of the water, but if danger threatens, they can dive. Rove beetles of the genus *Stenus* can dart across water by secreting a substance which lowers the surface tension behind, so that the beetle is shot forward by the higher surface tension of the water ahead.

Surface-dwelling insects with hydrofuge hairs are particularly vulnerable to rain storms, for if they are battered beneath the surface they get wet, and so drown.

Aerial acrobatics

Dragonflies, mayflies, stoneflies, alderflies, caddis flies and water beetles all have aquatic larval stages which undergo a spectacular metamorphosis to emerge as adults. As soon as the wings have dried, the insects can make their maiden flight. Most fly by day, but on bright moonlit nights, water beetles will migrate from one pond to another, and they can be fatally attracted to the moon's reflection in a glasshouse roof. The mayfly emerges as a subimago (the dun of fly fishermen), from which the imago (the spinner) hatches and lives for only a few days. The massed dance of male mayflies, on a warm, still evening in late May or early June, is one of the marvels of nature. The males collect in swarms, which tend to maintain their position, though individuals repeatedly fly up above the swarm and flutter gently downwards. This aggregation attracts a female mayfly which enters the swarm and ultimately mates with a male in mid-air. Both low temperature and heavy rain will inhibit the dancing, which is then postponed until conditions are more favourable.

Unlike most insects, dragonflies beat their front and hind wings independently of each other. Their excellent vision allows them to catch their prey in mid-flight. The males of the large hawker dragonflies spend much time flying back and forth defending their territory.

1 See how the surface tension keeps objects afloat by placing a piece of blotting paper on a bowl of water. Place a needle on top of the paper, which gradually takes up water and sinks, leaving the needle afloat, held up by the surface tension. Add detergent very carefully to the water and see what happens to the needle.

Experiment to show how the surface film will support the weight of a needle.

2 Collect a variety of pond life (one each of a water boatman, a flatworm, and a leech). Transfer them to an aquarium or pie dish filled with pond water and some weeds. Cover the top with a perforated cover. Place a transparent sandwich box or a glass aquarium on top of a sheet of graph paper on which the centimetre divisions have been numbered. Fill the container with pond water to a depth of 4 cm. Transfer each animal separately, and time how long each takes to move over a measured distance. Calculate their relative speeds.

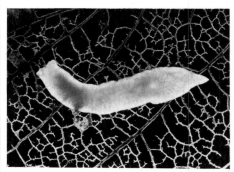

A freshwater flatworm *Dendrocoelum lacteum* crawls under water over a skeleton leaf.

3 Watch a hawker dragonfly and try to estimate the size of its territory. How many encounters does it make with other dragonflies?

A hawker dragonfly in mid flight defending its territory in August.

AMOROUS AMPHIBIANS

Amphibians are vertebrates with soft, unscaly skins which must be kept damp. While some newts spend all their life submerged in water, most amphibians return to water to breed and some have bizarre methods of reproduction.

Mainland Europe has a good collection of indigenous amphibians. But when the ice sheets melted after the last glaciation some 12,000 years ago, the sea level rose, flooding the landbridge connecting Europe to Britain and thereby preventing the spread of many animals—including amphibians—into Britain.

Air-breathing lungs enabled amphibians to extend the time they spent out of water and so be the first group of vertebrates to invade land. However, most modern amphibians are still unable to move far from water or moist surroundings. Three groups of amphibians exist today: the tailed amphibians, or urodeles (salamanders and newts), which are largely confined to the Northern Hemisphere; the anurans (frogs and toads), which have a world-wide distribution; and the worm-like caecilians, which are tropical.

When the air temperature starts to rise in spring, frogs and toads migrate back to their breeding pond, ditch, or canal (see pp.34-5). The males arrive before the females. One of the best ways of tracking down a breeding site is to listen for croaking males. Each species has its own distinct croak. Marsh frog calls become quite deafening when thousands of males gather together at one site. Some species, such as the common frog and the common toad, are active by day, and others, such as the natterjack toad and the European tree frog, are nocturnal, calling at night.

When a male frog or toad is touched, he gives a croak which repels an advancing male, but attracts a female. Once a male meets a female, he climbs on her back, grasping her tightly with his forelimbs beneath her belly. Common frogs remain paired in this embrace (in amplexus) until the female frog spawns. As the eggs are laid, the male fertilizes them externally before the gelatinous coating around each egg absorbs water and swells. Unlike frog spawn, which floats freely, toads lay their spawn as egg ropes which are entwined around plants under water.

▲European tree frogs *Hyla arborea* migrate into ponds to breed in May to June. During the rest of the year, they live out of water on reeds, bushes and trees.

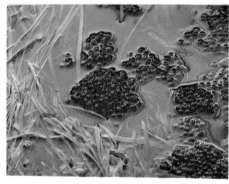

▶Frog spawn laid by the common frog *Rana temporaria* floats up to the surface after the albumen surrounding each egg absorbs water. Each female frog lays 2,000—4,000 eggs.

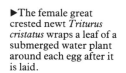

▲During courtship, the male pleurodele newt *Pleurodeles waltl* grips the female tightly above him by hooking his front legs over hers.

▶The female great crested newt *Triturus cristatus* wraps a leaf of a submerged water plant around each egg after it is laid.

◀The male midwife toad *Alytes obstetricans* carries the egg ropes, laid by his mate, entwined around his back legs until the tadpoles are ready to hatch.

▶A pair of common toads *Bufo bufo* in amplexus under water with the male on top of the larger female. She has just laid the egg ropes containing 3,000—6,000 black eggs

There is often a surplus of male frogs or toads, and then clusters of a dozen or so males can be found grasping a lone female, which may die from suffocation.

The midwife toad, which is found in south-west Europe, shows parental care of the eggs by the male. When the toads pair at night, the male winds the egg ropes around his hind legs as he fertilizes them. He then carries them around for 2 to 3 weeks, entering water at night to keep them moist and hiding by day. When the tadpoles are ready to hatch, he moves down to the water, to allow them to swim free.

Newts and salamanders do not attract a mate by calling; instead the male woos his mate either by chasing her on land (many salamanders) or by an underwater courtship (newts). Once a male smooth newt finds a female, he nudges his nose along her sides. After chasing her, he moves in front displaying his enlarged crest and bends his tail back along his body, fanning water towards her. The eggs are fertilized inside the female by the male dropping a spermatophore (sperm packet) which the female picks up with her cloaca; in other species fertilization occurs when the male and female cloacal openings are pressed against each other. In general, newts and salamanders lay eggs in water but some salamanders give birth to live young on land. Several species of newts lay eggs singly, wrapping a leaf around each one.

Compared with birds and mammals, cold-blooded amphibians and reptiles evoke less concern for their welfare among the general public. In Britain, only 3 of the 12 indigenous species of amphibians and reptiles are protected by law; yet France protects 52 of its 55 species. Europeans generally spend a great deal of money conserving their amphibians. In Germany, when toads are migrating, even main roads are closed to prevent massive road mortalities. Toad tunnels have been built beneath the autobahns. In Holland, portions of roads directly in line with toad migration routes are fenced with polythene sheeting and bucket traps are sunk into the ground. Regular patrols rescue the toads from the buckets and carry them across the road to safety. There is nothing organized on such a scale in Britain; although there are occasional signs warning motorists to take care.

Pollution of water by agricultural chemicals and the draining of ponds and infilling for building has removed many amphibian breeding sites and led to their decline. Fish introduced to ponds by anglers will eat large numbers of tadpoles, which means fewer frogs, toads and newts survive to breed another year.

PROJECTS

1 There is still much to be learnt about the breeding of even the common frog. Find out where the nearest breeding site occurs. If there is no place where frogs come, you can observe toads instead. Keep a record of the daily maximum and minimum air temperatures and the water temperature of the pond.

When do the first frogs (or toads) arrive?

Are they male or female?

How long is it before the other sex arrives?

How does the male, when holding the female in amplexus, keep other males away?

How close does a rival need to approach to elicit a response, and how does the rival try to dislodge the male in amplexus?

How long is it after pairing before the female spawns?

Is the spawn laid in shallow or deep water?

What happens to it when it is first laid?

How many eggs does a female lay?

How long does it take for the eggs to hatch?

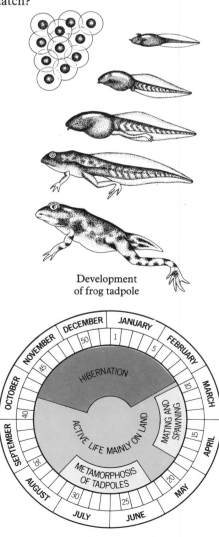

Development of frog tadpole

Year-cycle of common frog

PRODUCING OFFSPRING

The production of offspring is essential for the perpetuation of a species, and the timing is often linked to coincide with abundant food supplies.

Life within a pond reproduces in a variety of ways. Young may be produced asexually or sexually; they may hatch out from eggs or be born alive. Even after birth, there is much variation in the degree of care of the young (pp. 74–5).

Asexual budding

The simplest way of reproducing is for an individual to divide its body into two. The microscopic *Amoeba* does just this by constricting the body into a narrow waist producing two smaller amoebae that move away from one another. The flatworm *Stenostomum*, which lives in pond detritus, often reproduces by budding off a chain of two to four individuals. Flatworms in general, as well as sponges, can easily regenerate lost or damaged parts of their bodies.

When hydras have plenty of food, they develop buds, each of which has a free end with a central mouth surrounded by a crown of tentacles. The base then separates from the parent and a small replica is set free.

Parthenogenesis

Unfertilized eggs can give rise to offspring. This phenomenon is known as 'parthenogenesis', or 'virgin birth'. The best-known example is the ubiquitous terrestrial aphid. In fresh water, female water fleas will produce many generations of young during the summer months without any male being present. Nearly all rotifers, or wheel-animalcules, as they were known by early naturalists, are females which produce eggs that give rise to more females. As in water fleas, males are produced in unfavourable conditions and these fertilize eggs that subsequently develop a tough outer shell which can resist drought or freezing (pp. 20–1).

Sexual reproduction

Organisms which contain both male and female reproductive organs are known as 'hermaphrodites'. Leeches and some freshwater snails are hermaphrodites which come together

▼Nymph of the mayfly *Ephemera vulgata*.

▶The dun or subimago of the mayfly *Ephemera danica* hatches out from the nymph.

◀A pair of freshwater shrimps *Gammarus* sp in tandem, with the larger male carrying the smaller female beneath his body. She carries her eggs, and later her young, around with her.

▼A female fairy shrimp *Chirocephalus diaphanus* showing her brood pouch packed with eggs, which will survive long periods of drought when temporary pools dry up in summer.

◀The freshwater winkle or viviparous snail *Viviparus viviparus*, which is crawling over water mint, gives birth to live young.

▶A water flea *Daphnia obtusa* carrying eggs inside the brood pouch.

and mate, the male cells from one cross-fertilizing the female cells of the other, thereby ensuring offspring will be produced by both partners. Not all hermaphrodites mate, however; when hydras reproduce sexually, the male sperm are released into the water so they can swim to, and fertilize, the egg retained inside the hydra's body.

Species which have separate sexes must have some mechanism whereby a male and female are attracted to one another. The courtship dance of mayflies is described on pp. 24–5. The male white-legged damselfly has enlarged white hind legs which he dangles in front of the female before he carries her off in tandem. Male *Agrion* demoiselles open and close their brilliant metallic-coloured wings to court their female mates. The male damselflies take the female either in flight, or when she is perched, by grasping the top of her first thoracic segment with his anal claspers. The

large hawker dragonflies clasp the female by her neck; she then curves her body so the tip of her abdomen can collect the sperm from the male. During May and June the courtship of the swamp spider (pp. 24–5) can be seen in small, boggy pools. Outstretching his front pair of legs, the male spider vibrates them up and down on the water surface, out of step with one another.

Egg laying

After fertilization, eggs may be laid directly into the water (mayflies), on underwater vegetation (molluscs, caddis flies and water bugs) and on marginal emergent vegetation (alderfly); by dipping the tip of the abdomen into water (some dragonflies); and by insertion into plants under water (dragonflies and water beetles). Eggs may be laid singly (great water beetle); *en masse* in an egg cocoon (flatworms, leeches and oligochaete worms); or in

▼After pairing, the female of the caddis fly *Brachycentrus subnubilus* (or grannom) extrudes a green egg mass from her abdomen. During fine weather she flies upstream, dipping her abdomen below the water so that the egg mass (containing 60–700 eggs) breaks away. It then swells, becoming glutinous, and attaches itself to any underwater objects it meets downstream, such as these yellow flag leaf bases. In May, the grannom 'hatch' is a spectacular sight.

jelly (molluscs and caddis flies).

As in the marine environment (pp. 144–5), some freshwater organisms put their energy into producing large numbers of small eggs (the swan mussel produces half a million eggs), most of which do not survive; while others, such as the freshwater crayfish, lay far fewer, but larger eggs, a higher proportion of which survive. Thick-walled 'resting' eggs are produced by several freshwater animals as a means of survival during drought or severe winter weather.

Young stages

Not all animals discharge their eggs to the elements; some care for them by carrying them around in brood pouches until they hatch. Females of the crustacean *Cyclops* are instantly recognizable by the pair of egg sacs which project one from each side of the end of their bodies. Like all crustaceans, the eggs hatch into nauplius larvae, which undergo several moults before they emerge as a young *Cyclops*. Water fleas can often be found carrying eggs or young in their brood pouches. Female fairy shrimps carry their eggs in a single brood pouch which, since these animals perpetually swim upside down, appears uppermost. The water louse carries eggs and young in a chamber on the outside of its body. While female freshwater shrimps are carrying their eggs or young in their pouches, they are carried around in tandem by the males.

One of the most striking aspects of the life histories of pond inhabitants is among the insects, where a remarkable metamorphosis takes place from the larval to the adult stage. Young flies, dragonflies, mayflies, stoneflies, alderflies, moths and beetles all differ greatly from the adults. However, young springtails and bugs resemble small versions of the adult, except that juvenile bugs have no wings. They gradually change into the adult form by passing through a series of moults.

Viviparity

Another way in which eggs can be protected during their early development is by being retained inside the adult until the young are ready to hatch. While all mammals are live-bearers, relatively few invertebrates are viviparous. The freshwater winkles each produce about fifty young snails. The female *Cloeon dipterum*, one of the commonest pond mayflies, retains the eggs in the lower part of her egg-tubes for nearly two weeks, until she seeks water in which to deposit the newly hatched larvae.

Parasitism

Once eggs hatch, the young must find a source of food if they are to survive. Herbivorous animals, such as snails, lay their eggs on food plants, whereas carnivorous animals must go in search of their prey which often comprises a range of species. Parasites, on the other hand, are much more specific about their host species. The fertilized eggs of the swan mussel are retained in broad pouches for nine months until they develop into glochidia larvae, each of which has a pair of shells armed with teeth at its open end. To continue their development, these larvae must become parasites on fish. When a fish brushes past weeds with glochidia, the larvae stick to the skin and burrow down to form a cyst. After three months of living on the fish body, the mussel drops off to the bottom of the pond.

The oval red larvae of water mites are nearly always parasitic, usually on aquatic insects but sometimes on bivalve molluscs or even sponges. The water scorpion illustrated on p. 22 has been parasitized by larval water mites.

PROJECTS

1 Examine the undersides of waterlily pads and pondweed leaves for snail eggs. Collect one leaf and transfer it to a pie dish with water from the pond. Examine the egg mass with a hand lens. How many snails hatch out? How much of the leaf do they consume after one day, and after one week?

Egg ropes of the great pond snail *Lymnaea stagnalis* laid beneath pondweed leaves.

2 Collect fallen leaves with some water from a pond and put them in a pie dish overnight. Next morning, examine the surface with a hand lens for hydras. Transfer specimens to a jam jar and count them. With adhesive tape, fix a piece of black paper behind the jar and shine a torch from the side. Count the number of buds on each hydra. Add water fleas to the jar and observe at regular intervals for a few days. How long does it take for the hydras to bud off more individuals?

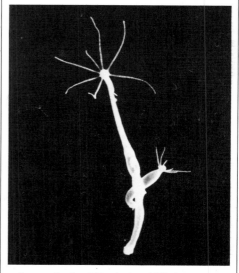

A green hydra *Chlorohydra viridissima* with a young bud.

3 Visit a pond in midsummer when damselflies and dragonflies are on the wing. Watch quietly to see how they lay their eggs. Do they hover over the water, dipping in their abdomens, or do they alight on a plant first?

DESIGNS FOR LIVING IN WATER

Living in water is very different
from living on land, and water
plants are adapted to these
special conditions
in a variety of ways.

In the atmosphere there is little
variation in light intensity and oxygen
concentration from the tallest tree to the
smallest moss. But in a pond there is
considerable difference in both of these
in the metre or so from the surface to
the bottom. Light intensity may be very
low on the bottom of a pond, especially
in full summer when the surface is
crowded with plants and plankton.
Oxygen diffuses 10,000 times faster in
air than in water, so is proportionately
lower even in the surface waters, and on
the bottom may be almost absent.

As all living tissues need to breathe
oxygen its availability affects the
distribution of many freshwater species.
However, all green plants produce
oxygen in daylight hours during
photosynthesis, so that, where light
intensity is sufficient for this process,
oxygen is produced. But even roots and
shoots in the mud at the bottom of the
pond need to breathe, so that air-
carrying tissue becomes an important
part of water plant adaptations, as does
increased leaf area in submerged plants
for catching light. All water plants are
buoyant with the gases that they
produce, so that they grow upright
supported by the water. A leaf that
appears to have a most particular shape
in water becomes a crumpled heap
when taken out of the water. Land
plants have to obtain their nutrients
from the soil, but water plants can
absorb them over the surface of the
plant from the water, so their roots
often have a rather different function,
as in floating plants, where they act as a
stabilizer.

Situations vacant
The surface The boundary between
water and air is a risky place to be,
where temperatures, levels and water
movement may vary from day to day as
well as from season to season. The
floating leaves of the waterlily are well
adapted to these conditions, with long,
slanting stalks that can alter their
position in the water, and a waxy upper
surface that resists wetting and
drowning. The breathing pores

Section of the solid
stem of lesser pond-
sedge *Carex acutiformis*
(**1**) with two enclosed
leaf bases

Section of a hollow
stem of the common
reed *Phragmites
australis* (**2**) and leaf
base wrapped round
the stem.

A section of a small
pond showing the
emergent, floating and
submerged zones of
vegetation and some of
their adaptions to living
in water.

Section of the stem of
yellow flag *Iris
pseudacorus* (**3**)
showing three folded
flattened leaf bases

Hairy aerial leaf (left)
and waxy floating leaf
(right) of amphibious
water bistort
Polygonum amphibium
(**4**)

▼*Vallisneria spiralis*, a
widespread tropical
plant used in fish tanks
and established on the
bottom of warm water
effluents from factories
in Europe, has a
remarkable method of
pollination. The female
flowers reach the
surface by a rapid
elongation of the stalk,
open, and expose the

fringed stigma. The
male flower buds are
released, float to the
surface, open three sail-
like calyx lobes and sail
across the surface with
the two stamens
projecting. After fertilization the female stalk
contracts in a spiral and
draws the fertilized
ovule to the bottom to
release the seeds.

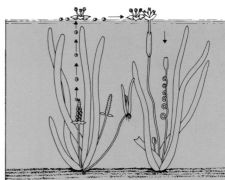

Branched bur reed
Sparganium erectum
(**5**): successive sections
through a leaf from the
submerged base to the
aerial blade.

Perfoliate pondweed
Potamogeton perfoliatus
(**6**)

The waxy floating
leaves and lettuce-like
submerged leaves of
the yellow waterlily
Nuphar lutea (**7**)

Hornwort
*Ceratophyllum
submersum* (**8**)

The floating rosettes of
frogbit *Hydrocharis
morsus-ranae* (**9**)

The strap-like
submerged leaves and
expanded aerial leaves
of water plantain
*Alisma plantago-
aquatica* (**10**)

Canadian pondweed
Elodea canadensis (**11**)

(stomata) are distributed on the upper surface instead of the lower, as they are in land plants. Amphibious bistort is a plant that is equally at home on land or in water, and changes its habit and leaf structure accordingly. On land it has an erect stem with stiff, hairy leaves that have their stomata mainly on the lower surface. In water it has a floating stem and soft leaves that are smooth and waxy above and have their stomata mainly on the upper surface. Both these plants are firmly anchored either to the bottom or to the bank.

Plants that are free floating on the surface must be adapted to maintaining their orientation at the surface as well as to surviving the cold weather in winter (p.34). If you look below the surface of the water at a frogbit or duckweed you will find the roots occupying a central stabilizing position in a rosette of leaves or a single frond, so that these plants are difficult to tip over. However, the surface is not a place to occupy in winter, and most of these plants have to have some alternative in order to survive it.

In the water Below the surface protection is afforded throughout the year, but this advantage may be lost in summer when the floating vegetation cuts down the light intensity. Some of these plants, such as lesser duckweed and hornwort, float freely in the water, but others are rooted with trailing, delicate leaves and stems, like the pondweeds.

Emergents These show the greatest variety both in adaptation to different conditions on the same plant and at different stages of their life-cycle. Most of these plants are adapted to supplying oxygen to their shoots and roots on the bottom of the pond, but this tissue is developed only where it is required. If you take successive sections through a shoot of branched bur reed you will find that the emergent leaf has a tough rigid structure with dense tissue that withstands the stress at the water surface. As you get lower down the tissue gets more spongy and full of air. The horsetails show increasing adaptation to a water habitat, tending towards a larger air space and a less rigid stem. Strength at the surface is achieved in a variety of ways: leaf sheaths are interlocked to make a narrow swordlike shape, in a triangle, or as a series of concentric cylinders.

Reproduction

Few water plants manage fertilization under water as most pollen is not resistant to water, so that many water plants rely mainly on vegetative reproduction, others emerge to flower and then return their seeds to the water (p.82). *Vallisneria spiralis* shows a series of remarkable adaptations to its still-water habitat. A widespread plant in the warmer parts of the world, it has spread into northern Europe where water is warmed by the effluent from mills. The very common Canadian pondweed has spread over Europe despite the fact that only the female plant was introduced at first and the male is still rare (pp.98–9).

Economic water plants

Although papyrus has long since disappeared from our paper-making resources, it is still frequently found in Egypt, the Middle East, and northwards to Sicily and south Italy. This handsome plant of still or slow-flowing waters is a sedge with a triangular stem that was used for making paper in ancient Egypt and Syria—hence the name paper.

Almost half the world's population is dependent on rice as a staple food. Originally native in temperate and tropical still waters, it is now widely cultivated in specially flooded fields, and is grown in the Po valley, the Rhone delta and in Portugal.

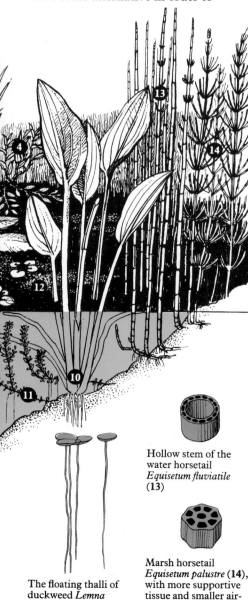

The floating thalli of duckweed *Lemna minor* (**12**)

Hollow stem of the water horsetail *Equisetum fluviatile* (**13**)

Marsh horsetail *Equisetum palustre* (**14**), with more supportive tissue and smaller air-spaces

1 Make a small pond, using thick polythene sheeting to contain the water. Surround it with turf so that there is an earth bank at the edge of the water, and put a substrate from a local stream—sand or gravel—on the bottom. (If you make an indoor pond nail the polythene to a frame, and use the substrate to build up the edges.) Introduce plants from the margins of local ponds. How do they behave, and at what rate do they spread? If you introduce submerged plants that are rooted at the bottom of the pond, do their leaves remain the same shape as they grow towards the surface? If you lower the water level what happens to the plants?

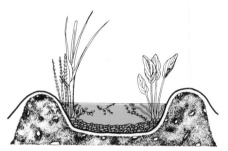

2 Collect some of the water plants in your area, and cut thin cross-sections with a sharp, single-edged razor blade to look at the pattern of air spaces. Put the sections on a piece of glass, and using a torch and a hand lens or low-power microscope investigate their structure by lighting them from above or below.

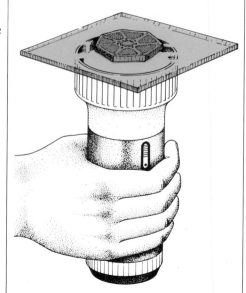

3 Collect seeds of water plants in the autumn and put them in jars of water in a cool place where you can observe what happens over the winter. How long do the seeds take to sink? When do they germinate? Do they all germinate at the same time?

A POND FOOD WEB

The whole pond is a delicate balance of the conditions required for each species to flourish, and who is eating what, and where.

Each freshwater body offers a range of physical conditions when it is first formed (pp. 16–17), but once living organisms have arrived many microhabitats develop around the plant species, which in turn affect the herbivore and predator populations. The basic conditions of water depth, basin shape and bottom material will vary over a small pond, suiting different species. At first a large number of only a few species colonize the pond, but as more arrive, competition and the development of microhabitats confine each species to the region where it performs best. But survival will also depend on the rate at which any organism is being eaten. If too many predators prefer one type of food, that food will disappear, and so will the predators, unless they find other food. Populations vary from season to season and from year to year, but the basic productivity of the pond is limited by the physical conditions of the area. The food producers are the plants, their productivity being controlled by the nutrient supply and the amount of sunlight that the pond receives. The number of herbivores and carnivores that the pond can support will depend on the amount of green material there is to eat. At each level energy is used in the pursuits of life—breathing, hunting for food, and breeding, hence the situation where a few top carnivores are dependent on many herbivores. The number of different species in the pond, however, will depend on the number of microhabitats that have developed over a long period of time. Thus the productivity of the pond does not significantly change, but the number of species and the complexity of the food web will increase with time.

Primary producers, herbivores and carnivores

At the production level, the pond separates into ecological units early in its history. The open water and surface zone supports the algae which float on the surface, while the shallow-water zone develops into reedswamp. Between these two communities there is usually a belt of floating vegetation which also has its own characteristic community. Herbivores are usually rather fussy creatures, preferring a diet of one species, so that they are frequently limited to one particular plant in the producer community. Their life-cycle stages must also be regulated to the period of maximum productivity and the dormant period of their food plant. This is a rather sedentary life, requiring neither the agility of a predator nor the means of finding and catching prey, so that herbivores are often fairly slow-moving creatures without conspicuous legs, eyes, and mouthparts. Carnivores, on the other hand, move around in search of their prey, requiring the ability to move fast, good sight, and efficient mouthparts with which to grab their prey. The smaller carnivores move around within a community, while the larger carnivores visit different communities for food within the pond. Near the top of the food web carnivores such as the heron may travel considerable distances, and visit several ponds for food.

▶A predator among the submerged water plants, this water scorpion *Nepa cinerea* has caught a damselfly naiad which it pierces and sucks out the body fluids.

A pond in summer, where production is at its height in the fringe. Floating and submerged plants ensure plenty of food for the many herbivores that graze here, as well as leftovers for the detritus feeders on the bottom of the pond. The smaller carnivores are found among the plants and detritus where prey is plentiful, but the larger carnivores, such as the trout, move over the water to catch a variety of organisms. Above the pond adult insects are on the wing, as well as several visiting birds that use the pond as a source of food.

The fringe and aerial zone: fringe species of monocotyledons including:
1. Common reed *Phragmites australis* (p)
2. Reedmace *Typha angustifolia* (p)
3. Yellow flag *Iris pseudacorus* (p)
4. Adult caddis fly LIMNEPHILIDAE
5. China mark moth *Nymphula* sp (h)
6. Reed warbler *Acrocephalus scirpaceus* (c)
7. Damselfly *Agrion virgo* (c)
8. Coot *Fulica atra* (o)
9. Dragonfly *Aeshna* sp (c)
10. Mallard *Anas platyrhynchos* (o)
11. Mayfly *Ephemera danica*
12. Heron *Ardea cinerea* (c)
13. Common frog *Rana temporaria* (c)
Surface and upper water zones:
14. Phytoplankton (p)
15. Waterlily *Nuphar lutea* (p)
16. Frogbit *Hydrocharis morsus-ranae* (p)
17. Duckweed *Lemna* spp (p)
18. Ramshorn snail *Planorbis planorbis* (h)
19. Wandering snail *Lymnaea peregra* (h)
20. Hydra (c)
21. Mayfly nymph *Ephemera danica* (h)
22. Water boatman *Notonecta glauca* (c)

A pond-scape

The open water In summer the successive blooms of algae are visible as a green haze at the surface (pp. 40–1), and there are many herbivores feeding among the algae, from the water fleas and rotifers (only just visible to the naked eye) to the larger mosquito and midge larvae. These small herbivores complete their life cycles very rapidly, so that when conditions are right for the algal plankton to grow the population of herbivores can explode very rapidly. The midge larvae have a method of

▶A predator on the bottom of the pond: the larva of the great diving beetle *Dytiscus marginalis* feeding on a tadpole.

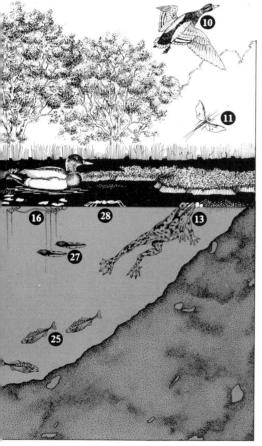

23 Great diving beetle *Dytiscus marginalis* (c)
24. Leech HIRUDIDAE (p)
25. Three-spined stickleback *Gasterosteus aculeatus* (c)
26. Brown trout *Salmo trutta* (c)
27. Tadpoles of common frog *Rana temporaria* (o)
28. Water measurer *Hydrometra stagnorum* (c)
On the bottom:
29. A submerged water plant and larva of great diving beetle *Dytiscus marginalis* (c)
30. Gnat larvae *Chironomus* sp (d)
31. Flatworm feeding on stickleback corpse (d)
32. Freshwater shrimp *Gammarus pulex* (o)
33. Caddis fly larvae in case LIMNEPHILIDAE (c)
34. Water hoglouse *Asellus aquaticus* (d)
key: (p) producer (plant) (h) herbivore
(c) carnivore (o) omnivore (d) detritus feeder

feeding that is often used by the smaller grazers; the bristles around the mouthparts agitate and create a current that wafts the algae towards the mouth. On the bottom the rain of dead algal cells and detritus feeds others, such as mayfly nymphs and a caddis *Leptocerus*, which constructs its case of sand. The fauna here must be able to live in low oxygen conditions and blood worms may be abundant here (pp. 84–5).

The floating plant or pondweed zone This zone is one of the richest in the pond in terms of species, and although the smaller organisms still occur, the larger ones are more conspicuous. The pond snails graze on the pondweed, and their jelly-like egg capsules are common on the undersides of the floating leaves. China mark moths betray their whereabouts by neat oval holes cut in the pondweed leaves; underneath the leaves, oval patches 'sewn' together conceal the caterpillars. A close look at the stems of the pondweed will reveal, anchored to them, an abundance of small plants and animals such as the hydra, which feeds by catching and paralysing prey with 'stinging' tentacles, then drawing it into the sack-like body chamber where it is slowly ingested. Where there is a wealth of herbivores there is also a wealth of larger carnivores, including the water boatman, great diving beetle and water scorpion. These fierce predators can move fast, and all have conspicuous grasping or piercing mouthparts to catch prey often several times larger than themselves. The larva of the great diving beetle is a bottom dweller and also a fierce carnivore, but must come to the surface for oxygen; other larvae may live among the pondweed that has submerged leaves producing ample oxygen all the year round.

The swamp zone—a sanitary service This is the zone of maximum productivity in the pond and all the aerial parts that are not consumed by grazers from the land collect on the bottom of the pond, making a layer of organic detritus. This is the feeding area for the thousands of cleaners, many of them invisible to the naked eye. More conspicuous are the vegetarian water hog louse and the flatworms that feed on dead or decaying animals by extruding a long feeding tube and sucking up the body fluids. They can scent the food from a considerable distance. A similar zone is created where trees overhang the pond.

In this small pond the top of the food chain is represented by the great diving beetle and the sticklebacks, but these may be taken by visiting carnivores, such as the heron. We consider these to be common, but when we count the numbers of organisms in the pond they are comparatively rare.

PROJECTS

① Tie a piece of raw liver on a string and submerge it in different areas of a pond, edge and centre, shady and open, for at least 15 minutes at each site. Where do you find the greatest concentration of flatworms? Collect a number of flatworms and devise an experiment to test their ability to scent the liver.

② Use a pond net to sample the different layers and zones of a pond. Sample the pond litter by slowly moving the net backwards and forwards, stirring up the litter. Pull the net out and allow the surplus water to drain off, then tip the contents into a white enamel tray and sort through the litter to find the invertebrate fauna. Place these in a jar of clean water and observe. Record whether they move fast or slowly, and have conspicuous eyes, jaws and limbs. Can you separate herbivores from carnivores? Take a sample from the open water zone, and from among any weed (also collect a piece of weed to obtain the fauna that are attached) and repeat your observation. Does the proportion of herbivores to carnivores vary in the areas you have investigated?

Please return your catches to the pond.

One of the many animals that uses the stems of water plants as a base for life, this hydra has just caught a water flea in its tentacles and drawn it into its sack-like body.

The great ramshorn snail *Planorbis corneus* grazing on the floating leaves of frogbit.

THE POND IN WINTER

The pond in winter looks inhospitable at the surface, but there is still plenty of activity lower down.

A temperate climate is characterized by a period of low temperature conditions unfavourable for growth and a warmer period when there is still an ample water supply. Temperatures below freezing damage plant and animal tissue, so that any species living in a temperate climate must be adapted to surviving the cold winter period. The surface of a pond in winter looks quite bare. Gone are the lush fringes of vegetation and the floating leaves of the pondweed and waterlilies, and only a few of last year's reed stems remain until broken by wind or rough water. Yet this apparently empty space is much frequented by visiting winter birds and also supports a resident population of coots and moorhens throughout the winter. These all feed here, providing they can keep a patch of water free of ice.

Water is a much poorer conductor of heat or cold than air, so that even in cold and rough weather, when there is considerable turbulence, the temperature of a pond falls slowly. Once a layer of ice has formed at the surface, the water below is sealed off from the colder atmosphere and from further loss of temperature, except in prolonged cold conditions. Ice does not impede the passage of light until it is more than 5 cm thick or has snow on it, so that, although the surface vegetation has all gone, the submerged vegetation still survives, and without the shading of the reedswamp or floating leaves may manage to photosynthesize in the winter sunlight. When the ice cuts the pond off from its atmospheric oxygen the presence of submerged vegetation may be a critical factor in supplying oxygen for the fauna of the pond. These submerged leaves often have a large surface area, either as the lettuce-like leaves of the waterlily or as the fine, feathery leaves of the water violet. These plants are rooted at the bottom of the pond so may produce leaves at any level between bottom and surface, but plants that are free-floating must use another device in order to get away from the inhospitable surface. The winter buds of frogbit are full of starch and without airspaces, and are thus denser than the parent tissue. They are

▲A pond in winter without the lush green fringe of summer vegetation. The mallard still find food in the rich organic mud on the pond bottom, where all the leaves off the trees support a community of litter feeders.

◄The ice at the surface does not prevent the water violet *Hottonia palustris* from photosynthesizing in the winter sunlight. The oxygen produced keeps the pond aerated when the ice cuts off the air supply.

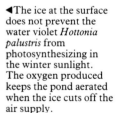

◄Frog spawn laid in a pond towards the end of the winter can stand being frozen for up to 12 hours without damaging the developing tadpoles.

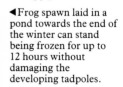

►Oligochaete worms living on the bottom of a pond are pink with haemoglobin that helps them obtain oxygen from water when there is little available.

easily dislodged from the floating plant, and, when they sink, remain upright on the bottom of the pond until spring. When the temperature rises the winter bud puts out slender leaves, using up the starch reserves and developing tissue with air spaces. The small plant rises to the surface and continues its

▲Mute swans *Cygnus olor* feeding on the bottom of an iced-up pond in a patch of clear water.

▼Changes in the shape of a water flea *Daphnia cucullata* during a year.

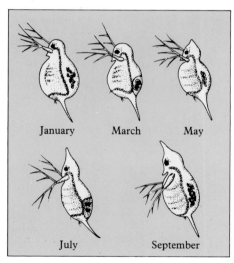

January   March   May

July   September

growth there. Duckweeds also produce dense winter thalli which sink to the bottom of the pond. Plants in the reedswamp zone store their reserves in underground stems and buds, and floating mats of reedmace will sink in winter. All remain dormant until the temperature rises in spring when the reserves are used up to develop the young shoots until they are able to photosynthesize.

As plants die back in the autumn, they produce a rich layer of organic matter on the bottom of the pond. Thus there is no shortage of food for the litter feeders, and many of the same invertebrate fauna are there. In the colder conditions their life processes slow down, so they require less oxygen and often move more slowly. The carnivorous larvae of dragonflies and other winged insects spend the winter on the bottom where there is a plentiful food supply. It is here that the ducks and swans feed. Some of the smaller members of the surface water community, such as the water flea, may develop special winter forms. Others enter a dormant phase and develop a resistant coat until conditions improve. The pond skater deserts the surface and spends the winter on land as a pupa.

Some of the larger animals like frogs, newts and toads, whose food supplies fall rather low in winter, hibernate. Frogs may hibernate in the mud at the bottom of a pond, although both frogs and newts will often leave the pond and bury themselves deep in leaves or mud. Frogs are able to absorb what little oxygen they need during hibernation through their skins. But winter is hardly over before they are making their way to the breeding ponds. Even in frosty weather frogs can be seen travelling overland in great numbers to lay their spawn in the 'right' pond. They usually travel upstream towards the pond and may return to the same pond for several years. We do not know what initiates this journey or how they find their way. Perhaps the scent of the water running out of a pond gives them a clue, but they can still find their way when water courses are dry. Both frogs and spawn are resistant to cold, and frog spawn can withstand freezing for up to 12 hours without damage to the developing tadpoles. One observer found frogs frozen in the ice and thawed them out: they hopped off quite unharmed.

The arrival of spring is accompanied by warmer weather and longer day length. Both these factors operate to break the dormancy of seeds and overwintering buds on the bottom of the pond and to stimulate the production of floating leaves in the pondweed zone, so that the pond may return to its summer state of productivity and population increase.

PROJECTS

① Collect duckweed in a floating state in the autumn. Put 100 thalli in a container outside and 100 thalli in a container on a windowsill inside. Record the temperature regularly, and observe how many thalli sink in each tank. Can you stimulate the floating thalli to bud by artificially increasing the day length with an electric light?

Duckweed *Lemna minor* thalli

② Collect some Canadian pondweed or water-milfoil in the late summer or autumn, and test what conditions stimulate the production of overwintering buds as in project 1. If you put some into distilled water—without nutrients—does this stimulate their production? (Obtain distilled water from any garage.)

x 3

Canadian pondweed
Elodea canadensis

③ Sample the bottom of a pond in winter, as you did for project 2 on p. 33. What invertebrates do you find? Is there a change in the proportion of herbivores to carnivores, or in population numbers?

TEMPORARY WATER WORLDS

Organisms that survive in temporary pools must either have a resting phase in their life cycle or spend some of it out of water.

Temporary pools are created every time it rains, in man-made environments as well as in natural ones. Blocked gutters or a hole in the road accumulate water just as do tree crotches or irregularly worn rock surfaces. These water bodies may last for a day to six months, according to conditions. Some appear regularly at a particular season, such as the turloughs of Ireland or the woodland pool illustrated here. These appear in the winter season when rainfall is high and evaporation or seepage rates slow. In temperate regions the rainy period usually coincides with the cold season, but in tropical regions it may coincide with the hot season, giving the hot, humid conditions of the monsoon. Where rivers are temporary features in the summer dry regions of the world, once the water has stopped running, the remaining pools may last many weeks, but organisms must be able to survive the dry summer. In limestone areas of the world, as in the Mediterranean, erosion through solution and evaporation gives a rough karst surface with many small but relatively deep seasonal pools that are teeming with life in the spring. In summer there is nothing to be seen but cracked mud on the bottom of the depressions.

A moment in time

A summer shower of rain fills a teasel leaf-base with water. Within a week this small pool contains a mass of chironomid larvae grazing on the algae on the surface of the plant. The following pupal stage is highly active, jerking up and down from bottom to surface in the small pool, until finally the winged adult emerges at the surface. These midges face the possibility that the pool may dry up before metamorphosis is completed, but have the advantage of living in a simple system that supports few or no predators.

A blocked gutter may produce a similar situation, except that the framework is more permanent than the teasel plant, and supports a plant population as well as the algae and

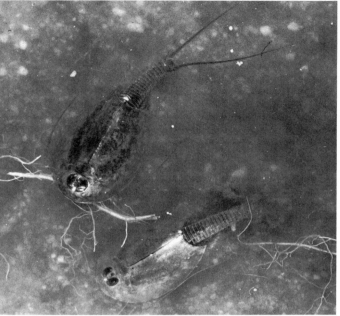

▲These primitive crustaceans *Triops cancriformis* feed in the mud at the bottom of a temporary pool. They occur frequently in the warmer Mediterranean areas of Europe, and although their eggs persist in a few temporary pools in Britain conditions are rarely ideal for them to hatch.

▶Water collects each winter in this depression in a deciduous woodland in south-east England, but dries out each summer.

▶Fairy shrimps *Chirocephalus diaphanus* live in temporary pools, perpetually swimming upside down with their undersides nearest to the light. By the time the pools dry out, the female fairy shrimps have produced their tough-coated eggs which survive throughout the summer in dried-up muddy depressions and hatch out when the winter rains fall.

▶A water bear *Tardigrada* is no more than 1 mm long, and is frequently found in gutters and temporary pools, where it can feed on the moss leaves. In dry conditions the animal can retract into a resistant cyst or tun, or survive as an egg.

▶The paired leaves of the teasel *Dipsacus fullonum* collect water in their connate (joined) leaf bases. This water supports a small population of midges, or chironomids, which as larvae and pupae graze on the algae growing on the leaf surfaces and then emerge as the winged adults shown here.

▼A blocked and rain-filled gutter has provided a habitat for some wind-dispersed plants of moss and willowherb. These then provide a source of food for other small organisms which will thrive until the gutter dries out in the sun, or is washed out in a rain storm.

Egg

leaves that accumulate in the hole. The decay of all this organic matter produces organic acids, giving the water a pH of around 4 that is tolerated by only a few organisms, including two species of mosquito.

A seasonal event

Pools that appear every year at a particular season may be quite extensive and productive water bodies, as are the turloughs that support a large population of winter wildfowl. Others, like the pools on limestone pavement, may be less than a metre across, yet teeming with life that miraculously appears with the pool.

Although a few of the plant and animal species may be invaders or opportunists (pp. 20–1), many of them have a life cycle that coincides with the life of the pool.

Organisms that cannot move away spend the dry season in a resistant resting phase. Fairy shrimps occur frequently in these pools in great numbers. Their nauplius larvae hatch out of the tough, drought-resistant egg case when the pool first fills with water. Both larva and adult feed by filtering organic particles out of the water. The crustacean *Triops cancriformis* also has a drought-resistant egg and is found widely on the continent, feeding on the detritus on the bottom of temporary pools. This species is common in the warm climate of southern Europe, but had not been found in Britain for 40 years when it was found in a Scottish pool in 1907. It has occurred sporadically since then in places as far apart as Kent and Gloucestershire, apparently surviving as an egg for many years until conditions are right for the adult to develop. In fact many of the smaller microscopic members of the temporary pool community, such as rotifers, algae and polyzoa, reproduce themselves in the form of statoblasts, which develop after desiccation or freezing. Copepods have appeared from mud that has been dry for ten years.

Water plants also appear as if by magic in seasonal pools, and many of these are annuals that spend the dry season as a seed with a thick coat which actually requires the sun to bake it before it can germinate in the next pool sequence. Some, like the versatile water crowfoots, seem to produce leaves for all occasions, including a dry land form that spreads across the mud of a dried-out pool.

In the woodland pool there is no sign of growing plant life, although there is abundant litter from the tree canopy above, providing the basis for a highly eutrophic pool which in February is full of cased caddis, chironomid and mosquito pupae. By June the pond is dry and the occupants have all emerged as winged adults and flown away.

chironomids. The plants are all wind dispersed, and while some, like the mosses or green algae, may stand the intermittent desiccation and continue to grow when conditions are right, the willowherbs were dead within a month of this photograph's being taken. Not so obvious are the minute grazers in this mossy habitat. Water bears or tardigrades are as adaptable as the mosses on which they feed, being able to pass the dry season in a dormant state or as a tough-coated egg. In this state they may be blown to other habitats. When the gutters are full they crawl in an ungainly manner among the moss leaves, feeding by piercing the surface of a cell and sucking out the contents. The real danger to organisms living in the gutter is of being washed out in a rain storm, but then there is always the chance of landing up in another suitable water habitat!

A water-filled tree crotch is like a mini-estuary, receiving all the nutrients from the bark in the runnels of water that collect here, together with the

PROJECTS

① Find some temporary water habitats in your area and observe them each week, recording their size, situation, depth and temperature, as well as finding out how many different kinds of organisms are living in each one. Collect and keep some of the small organisms in a jar to see how many stages in their life cycles there are, and how long it takes for them to emerge as adults. Make a chart to show the relationship between the age of the pool and the organisms in it.

The bole of this beech tree has a crotch that collects water running down the branches and trunk, and a supply of beech leaves. The water here is very acid, due to the decay of this organic matter, and is tolerated by only a few species of insect, such as the mosquito *Anopheles* spp and the beetle *Prionocyphon serricornis*.

A grassy hollow on a farm fills with water for short periods.

② Make up three experimental pools in polythene containers. Put about 800 grams of peat, garden earth and dried mud, from the bottom of a pond, into each container and add 2 litres of previously boiled and cooled water. Cover the tops with muslin to allow air to pass through, and keep for at least 10 days. What organisms can you find in each container after 10 days, and after 3 weeks?

LAKES AND MERES

Wind over Wastwater demonstrates the potential of wind power working across the 290 hectares of this lake, one of the largest and deepest of the lakes in the English Lake District.

Up to 78.6 metres deep, the lake affords plenty of room for summer stratification or layering to develop, where the warm surface water floats above the cold deeper waters until winter cold and storms cause the mixing of the water. It is this mixing that renews the nutrients in the surface waters to enable the floating phytoplankton to grow in the warmer summer period. These microscopic floating plant communities play an important role in the ecology of our larger water bodies as they provide the basic food material for the animal communities of the lake. This mountainous resistant rock landscape releases few nutrients to the waters eroding its surface; consequently the water is acid, or base poor, with a low pH. There are no settlements and no agricultural improvements to modify the oligotrophic ('few-feeding') nature of this lake. To compensate for the low level of nutrients the water carries little fine sediment, and light can reach the bottom over much of this clear lake, allowing some littoral species to grow on the bottom. On the edges, however, the waves may produce steep shingle beaches and erode any root-hold that a plant may obtain.

Wastwater, the deepest lake in the English Lake District, whose shores are exposed to the frequent battering of waves.

TROUBLED WATERS

Sudden 'water blooms' of microscopic algae occur in certain lakes, causing spectacular changes in the colour of the water.

Sunlight and wind are the two major physical factors affecting the water in a lake. Under their influence, lakes in temperate regions of the world undergo profound changes at different seasons of the year. In winter the water is cold and thoroughly mixed by wind-induced turbulence. During spring and summer the water near the surface becomes warmer; if the lake is relatively deep and sheltered, it may separate into distinct layers. This does not happen in wide, shallow lakes or small ponds.

Stratification of a deep lake forms three detectable layers: the lower layer, called the hypolimnion, retains the temperature of the water immediately before stratification (usually just below 12°C); the mid layer, called the thermocline, or metalimnion, has a steep temperature gradient of several degrees in a very short vertical distance; the upper layer, the epilimnion, exposed to the changing weather, warms up through the summer, developing a vertical temperature gradient in calm spells but mixing completely during windy periods. Because cold and warm water have different densities, the hypolimnion and the epilimnion become effectively separated, with the thermocline acting as an invisible barrier between them. Only when the surface temperature falls again can the autumn 'overturn' take place, fully mixing the waters of the lake once more and bringing fresh supplies of nutrients to the surface.

Colourful tales

There are legends associated with lakes in many parts of the world. Besides recording other less credible events, they often refer to sudden and 'miraculous' changes in the colour of the water. Only within the last century has it been observed that colour changes occur when there is a rapid increase in the number of microscopic water plants (algae) in the surface layer. The sudden onset of these 'water blooms', and the fact that neighbouring lakes may bloom at different times and with different species, remained a puzzle until recent work in Britain unravelled some of these mysteries.

Many kinds of algae and small

▶ Some typical phytoplankton algae and zooplankton grazers

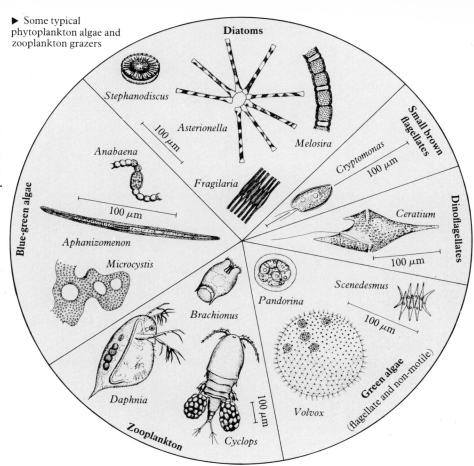

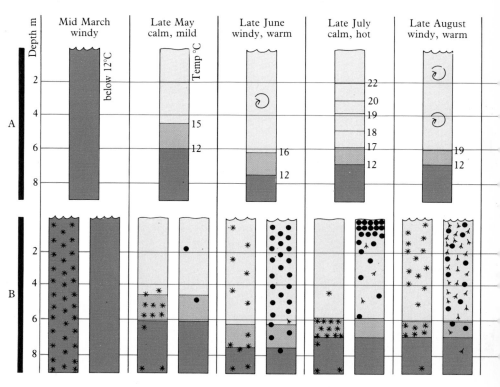

✳ Diatoms ⅄ *Ceratium* ● Blue-green algae

A. Imaginary columns of water from a mere in Shropshire, England, on seven sample days, showing temperature changes and the effects of turbulent mixing; the onset of stratification in spring; the three distinct layers on windy and calm days in summer; and the autumn overturn which restores complete mixing.

B. The vertical distribution of diatoms, blue-green algae and *Ceratium* throughout the year and in the weather conditions in A. The diatoms become 'trapped' in the thermocline in calm weather, while the blue-greens float to the surface and *Ceratium* remains below.

invertebrates live in the open water of lakes. They form the plankton, a free-floating community at the mercy of currents and eddies (the plants are called phytoplankton, the animals zooplankton). Water blooms are characteristic of meres, lakes and reservoirs that are eutrophic—that is, rich in nutrients, especially nitrogen and phosphorus. In winter the water is clear: a white disc lowered into the mere will still be visible at a depth of several metres; in summer, on the other hand, the water looks more turbid, and will hide the white disc at less than one metre's depth. Sometimes this turbidity is so thick as to resemble a green or brown snowstorm under the water. It is partly the result of the sun's rays warming and illuminating the epilimnion and so enabling the phytoplankton population to increase rapidly. As the tiny algal cells multiply—doubling every two to three days—they begin to consume the nutrients available in the water, to use up the dissolved oxygen, and to shade the cells below them like leaves in a forest.

Most of the planktonic animals either feed directly on algae (both living and dying) and bacteria, or prey upon other animals. Fast-growing algal crops can hold their own against grazing pressure.

Succession of species

The duration of a particular bloom is normally brief—days, not weeks—but each year may see a succession of stages involving different species. In the meres, diatoms (*Asterionella* and/or *Fragilaria*) dominate the early spring peak, and (mainly *Melosira*) take part in the late summer one. The same pattern on a subdued scale is shown in the minute flagellate *Cryptomonas*. The spherical green colonies of *Volvox* and its allies are commonest in the early summer, while the large brown *Ceratium* reaches an impressive peak in August or September. But the most conspicuous blooms are those of the blue-green algae in June and July—*Anabaena* or *Aphanizomenon* often followed by *Microcystis*: on still days they can form a stinking scum which paints the lake surface with bluish pigment as it decays.

Sink, swim or float?

How do the various species of the phytoplankton avoid sinking out of the sunlit surface layer? Three different mechanisms are involved:

The diatoms, each cell encased in a delicate glassy 'box' of silica, are actually denser than water. Like gliders or thistledown, they rely on an extended surface area to slow their sinking rate, and on upcurrents and turbulence to lift them towards the surface.

Cryptomonas, *Volvox* and *Ceratium* are self-propelled, like powered aircraft: their cells bear whip-like flagella enabling them to migrate slowly towards light.

The blue-green algae are passive but can become buoyant like balloons; their cells contain tiny 'gas-cylinders', or vacuoles—so small that an electron microscope is needed to see their structure—which are sometimes numerous enough to lift the algal colonies towards the surface in still weather. Their problem is the reverse of the diatoms': how to avoid over-exposure to the sun. The noxious surface bloom of blue-green algae is produced by flotation and death, not by population explosion.

It seems that the using up of nutrients is seldom the factor limiting algal population growth in a eutrophic mere. Light and temperature, the size and spatial geometry of the lake, and the sequence of still and turbulent conditions determine the dominant species at a particular place and time. Maybe the secret of controlling unwanted water blooms in reservoirs is to know precisely when to stir up trouble!

Figure

Late September windy, cool | Mid November windy

below 12°C

15
12

Lower layer, summer; whole lake, winter (less than 12°C)

Mid layer (thermocline) summer

Warm upper layer, summer

PROJECTS

1 Collect a sample of living plankton from a lake and examine it under the microscope to see what kinds of algae (and invertebrates) you have caught. If you have sampled a water bloom of blue-green algae, are they buoyant or not at the time of collection? Leave in the dark overnight: they should float to form a surface scum. If the cells are still healthy next morning, after about 4 hours in a well-lit room (avoid direct sunlight) their natural regulating mechanism will make them sink.

2 Fill a strong glass bottle with a sample of blue-green algae floating on water. Insert a tight-fitting cork to touch the surface—no air bubble. Hit the cork smartly with a mallet: the algal scum will slowly sink to the bottom, and will not rise again even if the cork is removed. The shock-wave has ruptured their 'gas cylinders'.

3 Small nutrient-rich pools—e.g. farm ponds—are often full of green flagellates and *Volvox*. Put a sample in a glass jar and completely enclose it in aluminium foil, except for a narrow slit. Shine a beam of bright light—sunlight is best—through the slit for at least 1 hour. Remove the foil, and see if the light beam has 'attracted' the motile green algae.

4 Dilute a sample of green algae and divide equally among 3 tall jars. Feed one jar with 3 or 4 drops of washing-up liquid, and one jar with a little nitrogenous fertilizer. Leave the other as a control. Leave on a sunny windowsill. What happens in 1 week, 2 weeks, a month?

Wind has concentrated the surface bloom of plankton along the water's edge of an inland pool.

FOOD FOR HOW MANY?

The animal population in every lake is ultimately dependent on the production by the plants which provide the base of the food pyramid.

Every green plant or algal cell uses the sun's energy to convert carbon dioxide and water into simple sugars; these form the building blocks for the carbohydrates, proteins and fats of which all living tissues are made. Plants also need basic nutrients, such as nitrates and phosphates, to grow. These are continually being cycled within the lake system, although the amount available for plant growth is partially dependent on the nutrients' entering a lake via the surrounding catchment area. All animals in the lake are totally dependent on the plant material produced in or around the lake or washed in by the rivers. Each animal uses up energy in its respiration and general metabolism, so only some of the energy from the plants is converted into body weight. This can be 30 to 60 per cent in the case of the water flea *Daphnia*, but as low as 7 to 10 per cent in adult fishes. Thus at each consumer level there is a large drop in the total body weight (or biomass) produced. The net result is that 1,000 kg of plant production may be needed to produce a single pike weighing 1 kg.

Variety in the lake

Plant production is dependent on sunlight and an adequate supply of nutrients. When conditions are optimum and there is little grazing by the animals, the phytoplankton community in the surface waters undergoes a population explosion, or bloom. Bloom conditions occur frequently in nutrient-rich lowland lakes and pools (pp. 40–1). The herbivorous animals increase their reproduction rate until either the plant growth slows because the nutrients are exhausted, or the grazing pressure becomes too intense. Herbivores have to be able to exploit food abundances, but still to be able to survive weeks or months of shortage, often either as eggs or in a resting stage. However, the large carnivores, such as fish, may take several years to reach adult size and so continually have to adapt their diet and

►A shoal of young minnow *Phoxinus phoxinus* that swim together as a body, making it hard for a predator to select one.

► The pike *Esox lucius* is a carnivore at the top of the food pyramid in the lake. A solitary predator of smaller fish, it requires a large territory to provide enough food.

▼A lake ecosystem showing the cycling of minerals in the lake, and the consumer levels or food pathways that are dependent on the producers in the phytoplankton or the flowering plants. When this is represented as a pyramid of productivity, a 1kg pike as a tertiary consumer is supported by 10,000 kg of producer plant material.

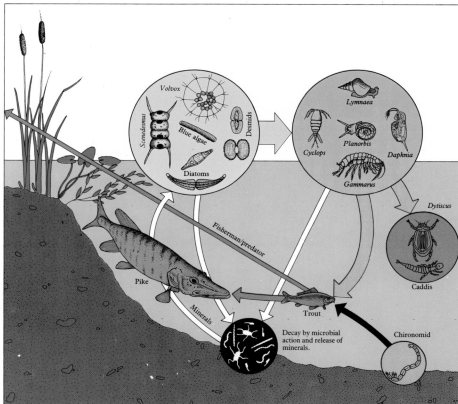

Schematic food pyramid

1 kg Tertiary consumers
10 kg Secondary consumers
100 kg Primary consumers
1,000 kg Producers

▼This flock of Barrow's goldeneye *Bucephala islandica* have come to Lake Mývatn in the desert-like centre of Iceland to breed and rear their young. This lake supports many thousands of breeding water birds for a short summer season. In winter ice seals the surface and the planktonic and invertebrate populations are either in a resting state or feeding in the rich organic mud on the bottom of the lake.

▼A female red-necked phalarope *Phalaropus lobatus* searches for food on the surface of a small lake in Iceland. Meanwhile the dull-coloured male is sitting on the clutch of eggs that she has laid, for in this bird the parental roles are reversed. In winter these birds leave the iced-up fresh waters for the shallow edges of the sea.

▲A rich green bloom of phytoplankton on an English mere in May produces the basis of a food web that supports the resident fish population and numerous visiting water fowl.

feeding ranges, both with the season and as their body size increases. The increase in size and weight ultimately depends on the amount of food available. This was demonstrated in the USA, where three similar artificial ponds were fertilized and stocked with 6,400, 3,200 and 1,300 bluegill respectively. At the end of the season the fish were removed and weighed, and it was found that the ponds contained 130, 120 and 122 kg of fish respectively, although the average weights of the individual fish were 15, 47 and 110 gm. The food had been shared out within the bluegill population of each pond. Therefore, in a lake which is artificially maintained with a single species of fish, the fewer the number of fish stocked, the larger will be their average size.

In nature, the food supply is shared by all the consumers at any level of the pyramid, but the numbers in each consumer level are kept in check by predation from the next consumer level. In a Swedish lake which had always produced plenty of large perch, all fishing was banned for ten years. The removal of man as a predator resulted in vast amounts of tiny perch in the lake and no big ones. Removing large numbers of the small fish restored the balance because the available food was shared between fewer fish.

This demonstration of how the food pyramid functions is often available for lakes where there is a fishing interest, and this same dynamic balance between supply, demand and predation pressure has also been demonstrated at lower nutritional levels where there are many more species sharing the available food. At times, the balance may be tipped by seasonal changes or freak weather conditions, so surpluses accumulate. These surpluses attract visiting predators, which move on again when the surplus is exhausted.

Winged visitors

Many water fowl travel enormous distances to feed or breed in particular places for a short season. Each spring in Mývatn, in the cold desert of central Iceland, thousands of ducks and other water birds arrive to rear their young on a lake which, during a short summer season, produces an abundance of food, mostly in the form of larvae and pupae of midges (the word *Mý* means 'midge'). The resident trout probably survive beneath the frozen winter surface by feeding on the invertebrates in the bottom muds, only to have to contend with competition for food when the migrant birds arrive. At the end of the summer, most of the birds travel far to the south to find winter quarters. The red-necked phalarope, however, takes to the sea in winter, where there is plenty of food to be found at the surface.

PROJECTS

1 Find out about fish catches in a lake or reservoir in your area. What type of fish are caught, what is the average weight of a fish, and how many are caught during each month of the fishing season? Are the fish fed? Is the lake stocked with fish each season? Which species are predators of the fish? Are these controlled?

2 Visit upland and lowland lakes where fishing provides an income. Does the average weight of a fish vary with the lake from which it comes? Test the pH of the water. Do you think there is any correlation between fish size and pH? What other factors could affect the size of the fish?

Wastwater	15.5 cm
Lough Atorick	17.7 cm
Windermere	21.5 cm
Loch Leven	28.2 cm
Malham Tarn	28.2 cm
Lough Derg	29.5 cm
Lough Rea	32.5 cm

Average lengths of trout in their third year, in a variety of lakes and tarns.

3 Sample the primary producers and the primary consumers in the surface waters of a lake in spring, summer, autumn and winter, by taking out a bottle full of water in a shallow area of the lake. What variety of types can you observe by using a x20 lens or a low-power microscope (pp. 40–1)? Sample the lake bottom in the shallows at similar times with a core sampler. When is there most plant detritus on the bottom of the lake? Does this coincide with high populations of consumers?

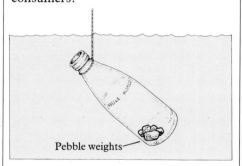

Pebble weights

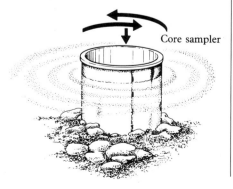

Core sampler

FISH POPULATIONS

Fish have long been important to man, both as a source of food and for sport. Their distribution and numbers depend on the environmental characteristics of the water they live in and their food supply.

Ever since man caught his first fish, he has been trying to find the best ways of catching more. Over the centuries, observation of life styles and feeding patterns has led to the devising of different methods for catching different fish. Effective lures have been designed and hooks appropriately baited.

To fish successfully, the fisherman must know where his quarry is and what it is doing. An interesting example of a highly varied fish community is provided by the great steppe lake of Neusiedlersee, in Austria. Here only one fish, the eel, is ubiquitous. The only other fish inhabiting the shallow fringes is the pond loach. In the reed beds live tench, crucian carp, pike, carp and bleak, of which only the latter two extend into the open waters of the lake. Restricted to the submerged weed zone beyond the reed beds are perch and roach, while rudd, bream, white bream and zander also occur there and in the open water beyond. Out in the open water the asp – a predatory fish that feeds on roach, bleak and small water fowl – occurs. This complex community of fishes includes species that feed on insects, snails and worms on the bottom, such as carp, tench, bream and pond loach; species that feed on insects, freshwater shrimps and snails in mid-water and on weeds, such as roach, rudd and bleak; and large predators, such as perch, eel, zander, pike and asp, that feed mostly on other fish.

As fish move around under water, they are aided by an additional sense of touch. The lateral line system, which is a canal lined with receptors extending along both sides of the body, senses low-frequency movements in the water. It helps the lone predator to detect the approach of the prey, and the member of a shoal both to keep stationary and to sense alarm movements by other members of the shoal that may be beyond their visual range.

European fish can be sorted into two fairly well defined groups: those that live in cool, well oxygenated, oligotrophic waters (pp. 38–9), which

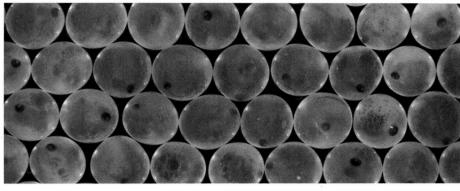

▲ Eyed rainbow trout eggs in a hatchery tray. Just before hatching the 'eye' is quite clear.

► Newly hatched alevins of brown trout *Salmo trutta*, with yolk sacs containing food reserves that will last two to three weeks until the fry are able to hunt their own food. The blood vessels of the yolk sac are visible, as well as the large oil globule which disperses into droplets as it is used up.

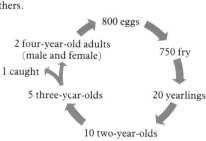

▲ These trout fry must now find their own food, and it is already obvious from their relative sizes that some are better at it than others.

► Trout fry at three months showing the characteristic 'parr marks' – the thumb-print bars on their sides.

800 eggs
↓
750 fry
↓
20 yearlings
↓
10 two-year-olds
↓
5 three-year-olds
↓
1 caught
↓
2 four-year-old adults (male and female)

▲ A simplified diagram of the survival of a trout brood.

► The bleak *Alburnus alburnus*, showing the line along its side which enables the fish to sense movements in the water around it.

are the game fish (many of which belong to the salmon family); and those which dwell in warmer, richer, eutrophic waters, like Neusiedlersee, which are the coarse fish. Both groups contain fish of considerable importance to man. The carp, for example, is a coarse fish which has been bred in fish ponds for many centuries, especially by monks, and carp ponds occur scattered throughout Europe. Game fish, particularly in northern countries, bring in a large revenue (the European average annual salmon catch is around 6,000 tonnes).

Game fish
Trout, as well as salmon, is an important game fish. Lake trout and the whitefishes perform migrations, like those of salmon, from the lake up into the head streams to spawn. These fish return to spawn in the same streams as they were spawned in, so even within a single lake system there are stocks of fish, spawned in different streams, that show slightly different genetic variations. On a larger scale, total genetic isolation of salmon stocks, after the retreat of the ice, produced distinct races in Sweden, North America and the USSR.

Trout spawning is initiated by a drop in the water temperature to around 6°C, and so, as is characteristic in the salmon family, takes place in winter. The spawning pattern is very similar to that of salmon (pp. 108–9). The female excavates a hollow, or redd, in the gravel bed of the stream, into which she lays the eggs, which are fertilized externally by the male. The eggs are covered with gravel and hatch out into alevins which do not feed but are nourished by their large yolk sac. When the yolk is used up, the young fish are known as 'fry', and they start to feed on small invertebrates. They develop a territorial habit, chasing off other fry that approach to within 10 cm of them. Fry with the best territories grow faster and then oust their neighbours. Eventually their diet switches to include bigger prey, such as minnows.

Coarse fish
There is a much greater variety of coarse than game fish in Europe, and there are interesting differences in their body structures and behaviour patterns, depending on the differing life styles. The fishes that feed on small organisms either in mid-water or on the bottom tend to form shoals. These help both to optimize the way in which the food supply is exploited and to give protection against attack by predators. Big predatory fish, such as pike, zander and asp, tend to be territorial and solitary. They often lurk, waiting for their prey to approach, before making a quick dash to seize it; whereas the shoaling fish tend to be continuously active, seeking out their food. Fishes that live in mid-water tend to have large eyes; whereas the bottom feeders, like loach, tench and carp, have small eyes and use the barbels around their mouth to detect their prey in the mud. Their colours vary according to their life style; the perch, with its vertical banding, usually lives among weeds; whereas the tench, a muddy bottom dweller, is a drab brown.

Lake management
Enrichment of a lake with sewage or fertilizer run-off from surrounding farmland will encourage coarse fish to increase at the expense of game fish. Management of lakes usually involves some effort to reduce the numbers of coarse fish. Coarse fish often spawn at quite high temperatures (10–20°C), and one female perch may lay up to 300,000 eggs, compared to a trout's 10,000. In one Swedish lake famous for its char (pp. 20–1), the numbers of perch are controlled by putting branches in the water at the perch spawning season; these are then periodically removed together with the egg masses.

Knowledge of the age structure of the stock is also important to the manager. Fish are aged by studying the rings, or circuli, on their scales. Most bony fishes have scales which grow by increments being added to the circumference (like rings in a tree trunk). Any check to growth, such as spawning, poor feeding in winter, or even pollution, will be expressed in narrower circuli, but usually a reasonably clear annual cycle is discernible. An alternative method, and the only method for scaleless fish, like eels, loach and wels, is to look at the otolith which also shows growth patterns.

To catch a fish
Today, detailed knowledge of the habits of fishes, their food, predators and life cycles is used to manage bodies of water to maintain good fishing stock. The greatest mystique is associated with fly fishing for salmon and trout, and the design of artificial dry and wet flies used to lure the fish into feeding, even when they are not hungry, is a great art. The fly is selected which most closely mimics the natural prey which is hatching at the time, for example, the dun or spinner of the mayfly in early summer (pp. 28–9). The coarse fisherman will usually tempt his fish with a baited hook near the bottom of the water, and will spend many hours in the same place, with another fisherman a few paces away. The habitats of these fish are often quite distinct, but where pollution, either by increasing organic matter or by warming of the water, occurs, coarse fish increase at the expense of game fish, and may have to be removed in large numbers.

PROJECTS

1 Find out which fish are caught in still and running waters near you. Are there any figures available from local water authorities or angling societies that enable you to obtain information on the populations of the fish in your area? What changes in population over the past 10 years, and in the past 50 years, can you find? Have changes in management initiated these? If stock is introduced, how does this affect the number of 'wild' species caught?

2 Obtain the stomachs of species caught by fishermen and examine the contents to see what these fish have been feeding on. Sort them out into plants, invertebrates, molluscs and fish. What proportion of these constitute each fish's diet?

Head of a predator – pike *Esox lucius*

3 Collect scales from some of the fish caught in the area, taking them from just behind the shoulder. Put these in an envelope labelled with date, species, size and location of capture. Clean a scale with a damp cloth and place it between glass photographic slide mounts, so that you can project it on a screen. Draw round the enlarged scale and count the number of circuli. Count the checks to growth. Record this information for different species and locations.

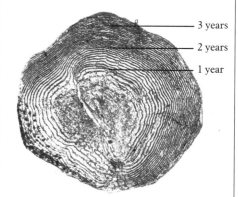
3 years
2 years
1 year

Scale from a three-year-old-trout that has grown up in good conditions where the circuli are only slightly reduced in winter, and no spawning has taken place.

EXTREME CONDITIONS

There are places in the world where conditions seem to be beyond the limits of tolerance for living organisms. Yet in perpetually hot springs or ice-filled lakes there are organisms that have adapted to these conditions.

There are many natural situations where conditions vary and appear to threaten life in fresh water. Water temperature at the earth's surface can vary from boiling point to freezing point, yet the thermal death point for most cold-blooded animals is between 30° and 40°C, and freezing destroys many living cells by rupturing their membranes. Growth may be impeded by too few nutrients or sometimes by too many. Naturally occurring minerals, such as lead and zinc, may poison the waters, and man has allowed these and many other poisons into freshwater habitats. One extreme condition may prevent the invasion of a habitat by the majority of species, even though all other conditions may be right; however, with time, an unusual assortment of tolerant species may survive.

Hot water/cold water

Hot-water springs and lakes are found in areas of the world where the earth's crust is fractured and volcanic activity frequent, so subterranean water is heated to great temperatures. At high temperatures salts and minerals in the rocks dissolve in the water, often making it blue. Once the water reaches the surface the temperature drops rapidly, but temperatures may still be around 70°C at source. At these temperatures only sulphur bacteria without chlorophyll can survive, but as soon as the temperature drops to around 50°C some species of chironomid larvae appear as well as algae, rotifers, beetle larvae and the wandering snail *Lymnaea peregra*. Once surface waters join the spring, the temperature drops still further and luxuriant growth occurs in the warm water. In cold regions, such as Iceland, hot springs will give rise to a luxuriant stream full of trailing water starworts and pondweeds and a fauna that is not found in neighbouring cold streams.

Ice lakes form where the barren

▼ This ice lake at the snout of the Breidamerkerjökull glacier in Iceland is rich in organic matter and nutrients from surface run-off and melt water from the glacier. Eider duck *Somateria mollissima* and red throated divers *Gavia stellata* are regular visitors here.

▶ Hot water springs at Ruaumoka's Throat in New Zealand are blue with silica dissolved from the earth's crust at high temperatures far below the surface. Only aquatic species that are tolerant of high temperatures can survive here but the warm soil and air encourage the growth of lush vegetation around the hot springs.

Europe during the retreat of the ice. Animals and plants that were abundant in these conditions are still found in alpine and upland regions, where they are relics from a colder climate. The flatworm *Planaria alpina* is found in alpine lakes and highland brooks where temperatures are below 11°C even in summer. It is also found occasionally in waters whose temperatures rise to 15°C in summer, but here it reproduces sexually in winter only when temperatures are low.

Nutrient conditions

Much more difficult to measure are the extreme conditions initiated by too little or too much organic nutrient. In upland regions iron-rich water depletes a lake of oxygen, and organic matter accumulates on the bottom as peat, giving a highly acid dystrophic lake. These waters are also low in other salts, such as those of calcium, magnesium and sodium, so no algae can live here. Plants that live on the edges of these lakes either are adapted to low oxygen concentrations and have 'air cells' to allow diffusion of oxygen to their roots, or, like marsh cinquefoil, may produce floating mats that obtain their oxygen at the surface. Chironomids still flourish there and these provide the food for the birds that visit and breed on these lakes in summer.

Where organic remains are abundant and readily recycled there is apparently no limit to potential growth, yet many species cannot tolerate eutrophic conditions and are eliminated when man adds organic matter from farms and sewage (pp. 84–5). These conditions do occur naturally, as in lowland regions, for example the Norfolk Broads in East Anglia, and lowland Holland. Calthorpe Broad was a highly eutrophic lake with a luxuriant macrophytic growth of many different species. In the summer of 1970, the water table in the Broad fell so low that formerly submerged peat was exposed. The iron compounds became oxidized and subsequently combined with the rain water to produce sulphuric acid and a deposit of iron hydroxide on the bottom of the lake. The pH fell to about 3, and many of the animals and plants disappeared. Since then the rain water has ameliorated the condition, and several species, such as the yellow waterlilies, have recolonized the Broad, but it will take many years for it to reach its former diversity. Man often creates these situations by using fresh water as a drain into which to pour unwanted chemicals, and if we do not know the previous conditions we cannot determine the losses. We do know that it was a combination of organic and chemical products which made a waste land of the Great Lakes, the Rhine and many other freshwater sites.

◀ Calthorpe Broad in East Anglia in 1980 – ten years after the drought that caused acute changes in the chemistry of this nutrient rich eutrophic lake with the consequent loss of many aquatic species. The yellow waterlilies *Nuphar lutea* are recolonising the open water but the lake will take many years to reach its former diversity.

▲ A dystrophic lake on valley peat near Mývatn in central Iceland where organic acids and low oxygen concentrations limit animal and plant life. Marsh cinquefoil *Comarum palustre* and sedges *Carex* spp. form a floating mat around the edges of the lake.

wastes of a glacier melt back from their terminal moraine, leaving a lake of 'melt water'. Large blocks of ice split off from the glacier and slowly melt in the lake. Surprisingly there is plenty of food here where the organic debris from the surface of the glacier accumulates and many birds, like the eider ducks shown, feed in these cold lakes.

Such conditions were prevalent over

PROJECTS

1 Collect a bucket of pond water that contains small organisms such as *Daphnia* and copepods. (You may not find these in winter conditions.) Take out several ½-litre samples in heat-tolerant containers (e.g. tins) and subject them to different temperature conditions: freezing, 40°, 50°, 60°C, and so on. Which organisms survive these conditions?

2 Treat two 1-litre samples from your pond water to different environmental conditions. Cover the containers with perforated polythene (to allow oxygen exchange but prevent evaporation) and put one container on a sunny window-ledge and the other in a permanently shady situation. Record the temperature variation in each water body. Leave for 2–3 weeks. Which organisms increase in number, and which disappear in the different conditions? What other factors can affect your results?

Sunlight

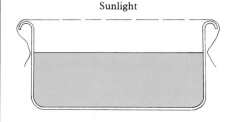

Shade

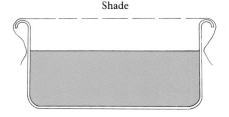

3 Choose some small ponds in your area with a variety of conditions, for example shaded and open, pasture, arable, moorland and farm pond. Record temperature, pH, substrate and environmental use, then sample your pond for free-living organisms. Can you correlate the species present with the conditions in the pond?

A heathland mire partially frozen over in winter

THE BIRTH AND DEATH OF A LAKE

A still lake may seem as much a feature of our landscapes as the hills around it. But how did it originally form, and how will it look in a few hundred years?

Once water arrives on the earth's surface as rain, it begins its journey to the sea, a journey of eroding the land surface and of carrying and depositing tons of sediment. If there is a depression in the earth's surface, or a layer of impervious rock, the water collects and forms a lake.

Streams entering a lake carry mineral and organic matter. When they reach the still waters they drop their load, beginning the long process of filling in the basin. They also carry nutrients from the surrounding land, and it is the amount of these that finally limits the vegetable and animal life of the lake. There is usually a variety of habitats offered to would-be colonizers: shallow to deep water, sheltered or open bays, rapidly silting areas to solid bottom; and this variety increases with the advent of the plants. The colonizers arrive mainly by air on the feet, bills, plumage or even in the droppings of waders and water birds. They might also travel overland on the fur of water mammals. Others arrive as seeds blown by the wind; true water plants are often inefficient at seed production, and spread mainly from small vegetative portions, often from other lakes higher in the system. In the early stages of colonization the reeds invade the margins of the lake to a depth of about one metre. Their underground rhizomes (stems) grow in a rapidly expanding network that stabilizes the sediment, while the tall vertical stems trap more sediment.

Each year, the leafy summer stems add more organic matter to the sediment, which in the de-oxygenated conditions on the bottom of the lake becomes reed peat. Further out in the exposed places, the hollow, leafless stems of the water horsetail grow up through about two metres of water. In sheltered places and deeper water, the floating, disc-like leaves of the yellow waterlily occupy another zone. Under the surface are its flimsy, lettuce-like leaves, produced mainly in winter to make use of the winter sunlight while avoiding ice or rough weather. In the

Rainfall catchment area

Pervious rock

Inflow

zone A

zone B

Impervious rock

1. A lake formed either at the junction of pervious with impervious rock, or by the temporary restriction of the downhill flow of surface water. Plants can colonize zone A, but not zone B where light does not reach the lake floor.

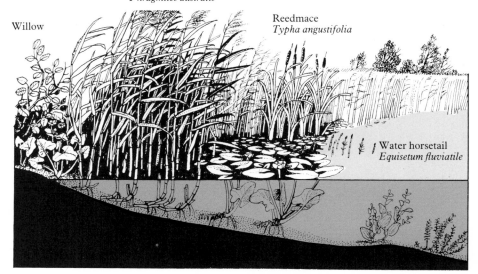

Willow

Common reed
Phragmites australis

Reedmace
Typha angustifolia

Water horsetail
Equisetum fluviatile

2. The colonization of open water by emergent, floating and submerged species of plants.

Yellow waterlily
Nuphar lutea

Perfoliate pondweed
Potamogeton perfoliatus

Canadian pondweed
Elodea canadensis

Willow and alder scrub

Tussock sedges
Carex spp

Common reed
Phragmites australis

Reedmace
Typha angustifolia

3. The lake basin is being filled in with sediment and reed peat. Tussock sedges are invading the shallow margins, followed by willow and alder scrub.

Yellow waterlily
Nuphar lutea

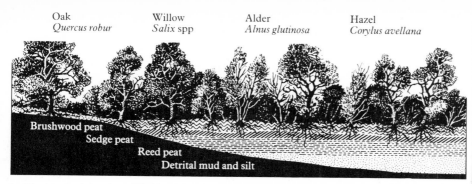

Oak
Quercus robur

Willow
Salix spp

Alder
Alnus glutinosa

Hazel
Corylus avellana

Brushwood peat
Sedge peat
Reed peat
Detrital mud and silt

4. The lake basin has been filled in with sediment and peat from different stages of colonization. The surface supports woodland, in which water loss through the leaves dries out the basin.

central area there is nothing to see at the surface; but there is a zone of submerged plants with translucent leaves through which water, gases and nutrients can pass. These plants need roots only for anchorage, and they readily break and spread by small vegetative portions. Their zone is restricted to the depth of water that allows the passage of light through it.

Many years on, the sediments have accumulated and the zones have slowly moved in towards the centre of the lake, displacing more water and filling in the basin. Each community has prepared the way for the next. Tussock sedges have moved into the reed swamp zone, providing places for marsh plants to colonize; and soon wind-blown seeds of willows and alders arrive and establish themselves on the tussocks above the water level, their shade gradually destroying the reeds. They also increase the rate of evaporation from the ground by their extra leaf area, so that there is a drying out of the surface.

With time and no interference, a woodland community may develop in lowland areas. If the rainfall is not high, the water table is controlled by ground water. The soil is aerated and drained by the deep roots and gradually dries out. There are hundreds of available living sites and nesting places, and many types of food in this wood, which will now be supporting many different species of plants and animals. These are interdependent, and are dispersed through the wood in a mosaic of

internally regulated conditions.

In areas of high rainfall, another community may develop on the surface of the sedge peat. Over the level surface of the infilled lake the rain water accumulates in pools that are invaded by sphagnum or bog moss. This small plant can cover a large surface area by building up a domed surface with a pattern of hummocks and hollows that retain the water against a developing gradient (pp.54-5). The surface is fed by rain water and is independent of the ground water below. In these conditions the sphagnum exchanges hydrogen ions for nutrients, and the release of the hydrogen ions increases the acidity of the bog so that only plants able to live in acid conditions, such as cotton grasses, heathers and whortleberries, can invade it. In lowland areas, birch and pine trees invade this acid habitat so that finally the area becomes a coniferous woodland of the kind found widely in high rainfall regions of the mainland of Europe.

What evidence is there for all this? Ever since the ice retreated from the European land surface, it has been recovering from all the alterations to its landscape. The evidence for these successive stages is written in layers of peat on the floors of old lakes that are now land surfaces. But part of the evidence is found in communities surviving on our wetlands today, although man has rarely allowed the land to reach the woodland stage.

1) Can you try and 'age' wetland plant communities, using the number of flowering plant species living together as an index?
1. Use a wire frame (called a quadrat), 25 cm square.
2. Choose an area where you know there has been some form of management, and the date when it last occurred, to compare with an area whose history you do not know.
3. Sample a small area, about 30 metres square, from each habitat, taking 20 samples from each.
4. Record each flowering plant, as it occurs in each quadrat. If you do not know it, make up a name and stick to it throughout the experiment.
5. Tick it as it occurs in successive quadrats.
6. When you have finished, your paper should look like this:

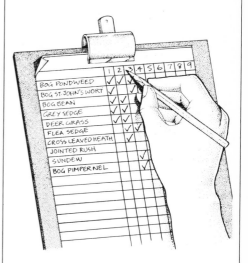

Then answer the following questions:
1. What are the dominant plants in your habitats?
2. Are they the same in each habitat?
3. How many plants are growing together in each quadrat? Work out an average.
4. Is the answer lower for the managed habitat? If not, try to find out what has happened in the unknown habitat to lower its species diversity.

2) If you know an area where bog moss is plentiful, collect a little and make a bog moss garden in a shallow trough with a glass cover to prevent evaporation. Water it with rain water, not tap water. If you have different types of bog moss (some are feathery, some are dense, some are yellow, green or red), make a scale map of their distribution, and their height in your trough when you put them in. Does this change in a few weeks? Try changing the water levels, and see what happens.

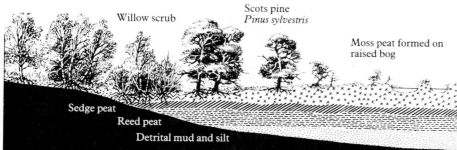

Downy birch
Betula pubescens

Willow scrub

Scots pine
Pinus sylvestris

Moss peat formed on raised bog

Sedge peat
Reed peat
Detrital mud and silt

5. A lake basin in a region of high rainfall that supports a raised bog above the ground water level, where only acid-loving heath plants can survive.

MIRES AND MOORS

The bright yellow-green patches of sphagnum moss on an upland moor warn the walker that a step in the wrong place will result in a bootful of ooze. A closer look at the moss reveals a mosaic of small moorland plants, which in late summer stud the moss with bright, starry flowers of butterwort, sundew, bog pimpernel and many others. Wetlands cover vast tracts of land in temperate uplands and northern parts of the globe, giving us the moors of Britain, the mires of Germany, the mosses of Scandinavia and the Arctic tundra. They have remained desolate, uninhabited regions, where there is hardly a living to be made except from summer grazing for sheep, cattle and ponies.

Today in our man-dominated landscape these areas are often greatly prized for their natural beauty, and for the many wild plants and animals living in them. Although we may enjoy the distant views of heather-clad hills, and the lark song bubbling over our heads on a summer's day, the harsh reality for the plant and animal inhabitants of the moor is the number of wet, cold and windy days occurring throughout the year. Sunlight is in short supply, as also are nutrients, which, instead of being released and recycled by death and decay, are locked up in waterlogged peat or carried to lowland areas.

A boggy flush on the Prescelli hills — part of the Pembrokeshire National Park created in Wales in 1952 — supports on its mossy surface a variety of unusual plants, including bog pimpernel *Anagallis tenella*, marsh pennywort *Hydrocotyle vulgaris*, and butterwort *Pinguicula vulgaris*.

PEAT AND PREHISTORY

Peat is removed to supply us with fuel and organic fertilizer, but it is also a record of past vegetation conditions and of man's activities before history was written.

► Garvin's track – a prehistoric trackway, exposed by present-day peat cutting, made from bundles of brushwood around 3,000 B.C. The birch and alder woodland behind develops on old peat cuttings.

Wherever there are waterlogged conditions and plants growing, the processes of decay are slowed down or arrested because few decomposers are able to live in anaerobic conditions (without oxygen). In these places plant remains may continue to accumulate as peat for centuries or millennia, until the water level falls, as a result either of improved climate or of a change in the drainage pattern. In some areas waterlogged conditions have allowed the formation of a continuous plant record from the time that the vegetation first colonized open water to the time that peat stopped forming (p.49). The nature of the peat will vary with the type of plants growing in the local conditions, but we distinguish two major types: moss peat, formed in acid water conditions, and sedge peat, formed in alkaline ground water conditions. But the evidence in the peat can tell us much more. It is a record of plant and animal communities that is laid down layer by layer—a diary of communities and events in the life of a bog. It is also a record of the impact of man on his environment, where the evidence may be preserved as man himself, his tools and buildings, or a change in the plant communities indicating management.

The story can only be revealed by excavation of the peat, layer by layer, down to the deposit on which it began to form. The evidence is contained in the peat itself but we now have many aids to the finding, interpretation and analysis of the peat remains. The macro fossils (seeds, roots, leaves and woody plant parts, insect exoskeletons and the bones, hair and so on of animal species) give very local and limited information, confined to the spot where the peat was extracted. A much wider picture of the vegetation of an area is obtained from microscopic analysis of the pollen contained in the peat. Pollen grains have a remarkably tough coat sculptured in a pattern that is characteristic of a genus and even of a species. Wind-pollinated trees, shrubs and herbs produce pollen in great abundance that is then dispersed over a

A hypothetical section through a column of peat on the Somerset Levels, in south-west England, showing the changing vegetation and habitat on the left, and on the right some of the artifacts that have been found in the peat.

With increasing rainfall a raised acid bog formed on the marsh dominated by sphagnum or bog moss, heather, crowberry, and cotton grass with birch and alder scrub at the edges. The forest is much reduced owing to the climate and also to clearance in the area.

With a rise in the ground water, fen conditions dominated the area with great saw sedge and abundant alders. There was a consequent decline in the elms and limes characteristic of drier conditions.

Freshwater conditions initiated the development of reed swamp in the basin.

400 BC	
Bog moss peat	
1400 BC	
Woody peat	
1600 BC	
Bog moss peat	
2200 BC	
Bog moss peat	
3500 BC	
Woody peat	
4500 BC	
Reed peat	
5000 BC	
Blue clay	

The Iron Age lake village contained many artefacts of iron, bone and wood.

Meare heath trackway, made of stout, pegged planks cut with bronze axes.

A wooden hay or brushwood fork.

Flint is not found in the area. These flakes were packed together in moss, probably in a bag that has perished.

Hurdles were widely used to make tracks across the wetter areas. They were constructed of coppiced hazel shoots.

The Sweet track is a planked walkway across the bog. Jadeite and flint axes, and a broken pot with hazelnuts spilled around it, were found beside the track.

► A diagram showing amounts of pollen of tree and herb species found in a section of peat on the Somerset levels, indicating the changes in composition of the woodland and pasture occurring with man's arrival in the area.

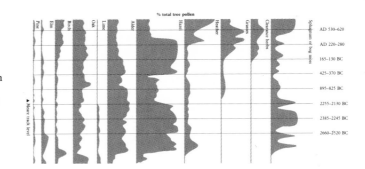

large area (about 150 km in any direction). If pollen samples are taken throughout the deposit, and the varying amounts of pollen produced by each species compensated for, it is possible to build up a picture of long-term change in the vegetation of the area. Thus agricultural man's arrival coincides with a fall in the amount of tree pollen and a corresponding rise in the amount of herb or weed pollen.

The dating of a peat deposit had long presented problems as, although each deposit is laid down sequentially, unless there was some datable clue, like a trace of human occupation, it was impossible to tell when the sequence began or ended. Since 1949 we have been able to reconstruct the time scale backwards many thousands of years by using radioactive tracers. Radioactive carbon is present in small but fairly constant amounts in the atmosphere as carbon dioxide, which is regularly incorporated into plant tissues until the moment of death. After death the radioactive carbon decays or loses its radioactivity at a slow but constant rate, taking about 5,370 years to reach its half-life. By measuring the amount of radioactive carbon left in a plant or animal (plant material being incorporated as food) sample, we can estimate the length of time that has elapsed since death. In older deposits, where there is little radioactivity left, inaccuracies creep in, and the ± figure after the date gives you the range of accuracy of the dating of the sample. Where forest cover has been continuous, tree ring dating (dendrochronology) can be used. Each year a tree adds a ring of tissue to its trunk, whose radius will vary with the growing conditions of that year. Seasonal variation is usually relatively consistent in an area, thus the tree rings can be matched and the time scale extended backwards as long as there are no breaks in the sequence.

A story in peat on Sedgemoor

This area of green meadows divided by ditches full of yellow irises ceased to accumulate peat about 1,300 years ago. Climatic changes are reinforced by drainage and constant removal of surface water. Peat began to form about 6,500 years ago, when the area was flooded after a rise in sea level following the retreat of the ice.

A basinful of peat is a valuable commodity as well as a source of history. In this part of Sedgemoor in south-west England the giant peat extractors eat through thousands of years of evidence each day. Yet the story would not be uncovered without the co-operation of the men who drive these machines and their names are commemorated in many of the archaeological finds.

There were two Iron Age lake villages in this region of Sedgemoor c. 400 BC. Meare lake village seems to have been a regularly used summer camp of about 37 clay floors surrounded by slanting posts that probably supported skin tents. The floors are laid on timber and brushwood directly on the surface of a raised moss bog—a very wet place to choose to come and live each summer. The rubbish tips give us a clue. Excavation revealed 39 species of water fowl, sea birds and birds of prey that were caught in the area, as well as domestic animals and crops. The list includes white-tailed sea eagles, kites, peregrine falcons, gannets, goosanders, red-throated divers, and many other now rare birds. To obtain these wild fowl the villagers had to venture from the camp across the marshes, a region of bogs and forest where it was difficult to find their way. So they built timber trackways, made of split oak laid across timber 'sleepers'.

In other places hurdles or bundles of brushwood were laid across the marshes. This tradition had already been in existence for some time. The hurdles were easy to construct, light to carry and remarkably firm on the bog surface. The Bronze Age hurdle trackways are distinguished by the axe heads occasionally found lost in the bog beside the track.

The earliest trackway of all is also a timber trackway: the Sweet track, named after Ray Sweet, the peat digger who first saw it. It is hard to believe that Stone Age axes felled and split the stout oak timber of which this track is made; but the tools are there beside the track. There are wooden mallets, stone axes, often from far distant places, wooden food dishes, and a poignant reminder of an October mishap around 6,000 years ago: a broken pot with spilled hazelnuts around it.

The vegetation evidence continues into the distant past, before man used the area, through the forest period to the scrub that invaded the sedge peat and back to the reeds that colonized the open water on the estuarine clay (pp.48–9): a story that can be found in many of our European peat deposits but which always has a local flavour.

The sphagnum story

Sphagnum or bog moss is both the main constituent and the cause of vast areas of upland bogs and mires in Europe. Its ability to retain water on a slope is demonstrated in this aerial photograph of Claish Moss in Scotland. The ground slopes towards the lake on the left of the picture. The slope is patterned with the hummocks of sphagnum and the water-filled hollows.

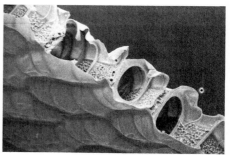

When you squeeze a handful of bog moss, the water pours out, but when you release the pressure the moss springs back into shape, and, if it is near a water source, will reabsorb almost the same quantity of water. Yet there is no water-carrying tissue in a moss shoot.

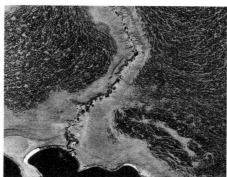

A sphagnum hummock begins in the water with a floating raft of fine-leaved bog moss that is rapidly colonized by the hummock forming species *Sphagnum papillosum*. Other species move in, and the hummock becomes a mosiac of species each adapted to slightly different conditions.

This scanning electron micrograph shows the inside of the cells of a leaf of *Sphagnum papillosum*. The water-retaining cells are spirally thickened and run in a continuous interconnecting network through the leaf, aiding rapid absorption of water and retaining the cell shape and water content in adverse conditions.

WETLANDS

Wetlands form where water is held up on its journey from mountain to sea. They are characterized by large areas of bog and mire communities where man has hardly interfered with plant and animal life.

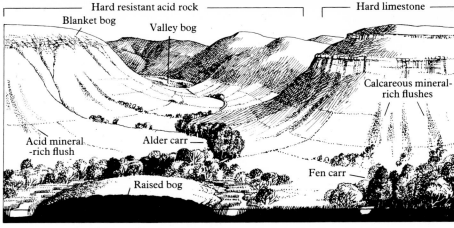

Hard resistant acid rock
Blanket bog
Valley bog
Hard limestone
Calcareous mineral-rich flushes
Acid mineral-rich flush
Alder carr
Fen carr
Raised bog

Over 230 million hectares of the earth's surface are covered by wetlands, where climate and geography have contrived to create many different kinds of permanently wet habitats from mountain blanket bog to valley mires and fen carr. The plants of these habitats are adapted to the waterlogged and climatic conditions in each situation and produce a characteristic landscape that we associate with our bogs and mires.

In each case the type of bog or mire that is formed is influenced by the climate, the geographical location and the rock type of the area. In flat highland areas where the rock is hard and impermeable, and the rainfall over 200 mm a year, large areas of blanket bog may develop. The main constituents of these bogs are the bog mosses, or sphagnum species. These mosses can retain water even on a slope, and build hummocks so that the bog surface appears rather like washing in a basin with the main bulk below the surface but some always visible above (p. 53). However, the depth of the layer of moss peat that can develop is limited by the slope and the tendency for these bogs to creep downhill, so that although peat-forming conditions exist it never reaches any depth except in hollows. Sphagnum further alters the environment for other species as it actively exchanges nutrients for hydrogen ions, which are then released into the bog, further increasing its acidity.

Few species can live here other than wiry-leaved sedges and rushes, and dwarf shrubs with small tough evergreen leaves that are resistant to cold and evaporation. Animal life is also reduced in a blanket bog, for the water is very low in oxygen as well as subject to extremes of temperature.

Valley bogs on resistant rocks are still acid but they are already carrying organic material from the vegetation above as well as any sediment and minerals that have been eroded from the bedrock. These accumulate in the more sheltered valley bottoms, where a rather different flora can develop.

▲ A diagram showing the relative positions and origins of different types of wetlands.

▶Alder carr, a nutrient-rich and shady wetland wood that supports early flowering plants of yellow flag *Iris pseudacorus* and marsh marigold *Caltha palustris* on the marshy area between the trees.

▼Fen meadow that has been recently cut for hay showing a rich variety of plants in the meadow and at the edges. Purple loosestrife *Lythrum salicaria*, milk parsley *Peucedanum palustre* and common valerian *Valeriana officinalis* thrive in the meadow, and bog myrtle *Myrica gale* survives on the margins of the woodland beyond.

▶Blanket bog formed on a highland plateau. Pools of water are invaded by the pale bog moss *Sphagnum* forming hummocks which are colonized by cotton grass *Eriophorum angustifolium* and deer sedge *Trichophorum cespitosum*. Older and larger hummocks support woody shrubs of heather *Calluna vulgaris* and cross-leaved heath *Erica tetralix*.

▼A valley bog in the Glencoe pass, Scotland, where the accumulation of minerals and nutrients supports a rich flora that includes devil's bit scabious *Succisa pratensis*, cross-leaved heath *Erica tetralix* and bog myrtle *Myrica gale*.

Sphagnum moss still covers the surface but here and there are delicate trails of bog pimpernel and marsh pennywort, and in summer yellow spikes of bog asphodel along with bog bean and a variety of rushes and sedges. Where grazing is prevented, willows can arrive. In this situation peat can form quickly, except where the water drains out of the bog, and may even rise up the sides of the basin or valley. As farmland appears the nutrient load of an acid hill stream goes up. Its sediment load is also greater and wherever the stream is slowed down it drops its load. Tree species colonize the marshy area, in the improved climate of the lowland region,

and further impede the drainage of the water. The end result of this is a boggy woodland of alder carr with pools of open water and a dense ground layer of shade-tolerant species of flowering plants. Alder carr was once widely established all over Europe but is now scarce owing to agricultural reclamations, and usually confined to stream edges.

There are several situations in which acid bogs more typical of highland areas may develop in the lowlands. They may be found in large areas of nutrient-poor sands, blown over impervious clays that are largely fed by rain water. Here they are characterized by similar species to the blanket bog, with large numbers of carnivorous sundews and butterworts that supplement their frugal nitrogen diet by catching insects (pp. 58–9). In low-lying regions of high rainfall an acid raised bog may develop on top of an alkaline fen. Here it is fed solely by rain water which together with the bog moss makes a very acid habitat only colonized by plants tolerant of these conditions.

Wetlands formed in calcareous or alkaline ground water will be very different from acid bogs, as the water provides a nutrient-rich basis for life, and the fenlands formed in these waters contain a great variety of plants and animals. They begin life with reedswamp colonizing the open water. The peat accumulates fast, filling in the basin until the area, if left undisturbed, becomes a rich mosaic of fen plants and animals with a high proportion of woody and tree species. Man has long used these areas as a source of wood, and managed them as hay meadows or for grazing his stock. These old meadows still contain many unusual fenland plants and animals that are adapted to the hay cutting or grazing routine. The ditches are full of species that would have been found in the fen pools. More recently the agricultural value of the nutrients locked up in the fen peat has led to full-scale reclamation and drainage or to excavation of many of our fenlands, so that a fen meadow, like the one illustrated, is a scarce habitat today containing some of our rare and disappearing species.

Where the mineral-rich waters appear in exposed highland areas as flushes, they are distinguished by the bright green carpet of vegetation and the variety of plant species that include some of our smaller and rarest montane species.

As a result of increasing pressure for productive land many of these wetland habitats are fast disappearing to farmland, or are so altered by drainage and fertilization of the surrounding land that many of the specialized plants and animals of these environments are finding it increasingly difficult to survive.

PROJECTS

1 Find out about your local wetlands: (a) history: look at old maps, available at local record offices, to see if areas of wetlands have increased or decreased in the last century. There should be late nineteenth-century maps marking wetlands available, such as the first medium scaled Ordnance Survey maps in Britain. (b) present status: look for their present flora and any indication of past management such as grazing, drainage channels, and their maintenance, tree stumps, etc. Contact your local Natural History Society or Trust for Nature Conservation to see if there are any records available. Can you help expand these?

2 Make your own rain gauge to measure the rainfall variation in local areas. Compare your results with those of the local meteorological station. Record soil water level by digging a small pit to ground water level and inserting a permanent post or metre rule. Record weekly or monthly readings. In which months is it highest/lowest? Compare the results with local rainfall figures.

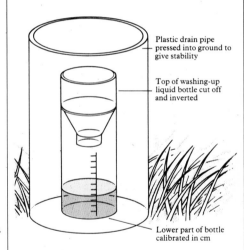

Plastic drain pipe pressed into ground to give stability

Top of washing-up liquid bottle cut off and inverted

Lower part of bottle calibrated in cm

Reading gives cms of water deposited over equivalent area in *y* hours.

3 Try measuring the rate of evaporation from an open surface of water in different weather conditions and situations. Use standard sized plastic pots, e.g. margarine or cheese containers around 10–15 cm in diameter. Fill each pot with the same quantity of water from a measuring cylinder and place them in a variety of situations (high/low; open/shady; north/south) for at least half a day. Measure the amount of water left and work out the rate of evaporation per hour. Repeat in different weather conditions. Which conditions cause the most evaporation?

CONSERVATION OF WETLANDS

With man's ever increasing demands for land, it becomes both more urgent and more difficult to evaluate our needs and to decide how we can attempt to conserve our wetland wildlife.

With escalating population figures, the demand for productive land goes up and up. Wetlands have for centuries yielded useful products for man, such as reed and sedge for thatching, peat for fuel, and rich summer grazing in meadows flooded in winter. Today, these products have been largely superseded by man-made ones, but peat is removed in large quantities for horticulture (pp. 176–7), and the meadows are drained to increase productivity. This increasing management usually means the destruction of the wild habitat; yet parallel with man's destruction has come a growing interest in the countryside and concern for our disappearing native species of animals and plants, so that many of these are now in specially managed reserves. But good management needs an understanding of the physical conditions of the habitat, as well as of the requirements of the species that inhabit it.

A boggy alder carr is not an easy place to walk through, and nor is a reed swamp; because of this many people regard these as desolate, inaccessible places that we could easily do without. Yet for the naturalist these and many other wetland areas provide unending opportunities to observe and learn about the interrelationship between species and their habitat. The knowledge acquired is an essential part of our understanding of an environment as an ecological unit, in terms of the space and conditions needed for survival. Armed with this information, we can evaluate and compare its worth as a wildlife reserve with its worth as a man-supporting unit. We have lost many sites to farmland, reservoirs and plantations through lack of proper evaluation.

Filling in the picture

A detailed study of any habitat reveals the energy budget between producers and consumers. This is relatively simple in an arable field where a single crop is

▶A black-tailed godwit *Limosa limosa* approaches its nest in a damp meadow. This species was formerly common across central Europe in damp hay meadows and wetlands, but is now restricted to wetlands or more northerly meadows where the mowing is late.

▼Cutting reeds *Phragmites australis* on the reserve at Wicken Fen in March maintains an open reed bed succession and a crop for the National Trust, which owns the reserve.

▶The swallowtail butterfly *Papilio machaon*, here resting on hawthorn, lays its eggs on the rare milk parsley *Peucedanum palustre*, on which the young caterpillars feed. The plant survives among the mown reed and sedge beds.

▼Conservation volunteers clearing a pond that has become dominated by yellow waterlily *Nuphar lutea*, in order to maintain some open water habitats and diversity of cover.

◀Exploitation of the vertical regions of reed stems *Phragmites australis* by different species of bird during the winter and summer months.
1. Bearded tit *Panurus biarmicus*; 2. Blue tit *Parus caeruleus*; 3. Penduline tit *Remiz pendulinus*; 4. Reed bunting *Emberiza schoeniclus*; 10. Wren *Troglodytes troglodytes*; 6. Moustached warbler *Acrocephalus melanopogon*; 7. Great reed warbler *Acrocephalus arundinaceus*; 8. Reed warbler *Acrocephalus scirpaceus*.

Winter　　Summer

▼Distribution of some characteristic birds in the reed fringes around Lake Neusiedlersee, in a section from open water to willow carr.
1. Moorhen *Gallinula chloropus*; 2. Coot *Fulica atra*; 3. Water rail *Rallus aquaticus*; 4. Spotted crake *Porzana porzana*; 5. Little crake *Porzana parva*; 6. Moustached warbler *Acrocephalus melanopogon*; 7. Sedge warbler *Acrocephalus schoenobaenus*; 8. Bluethroat *Luscinia svecica cyanecula*; 9. Reed bunting *Emberiza schoeniclus*; 10. Great reed warbler *Acrocephalus arundinaceus*; 11. Bearded tit *Panurus biarmicus*; 12. Reed warbler *Acrocephalus scirpaceus*.

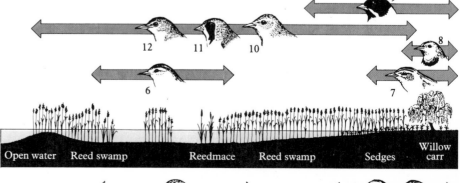

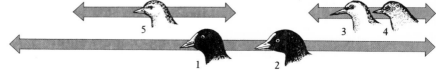

Open water　Reed swamp　Reedmace　Reed swamp　Sedges　Willow carr

grown and removed, and every other species in the habitat having also to adapt to the managed conditions. In a semi-natural environment it may take many years of patient observation to fill in the picture of the diurnal, seasonal and annual patterns of relationships between the species living there.

The Neusiedlersee and associated pools have been studied intensively for many years now by a team of scientists. This vast, shallow, inland lake on the borders of Austria and Hungary, 25 km long and up to 7 km broad, is surrounded by extensive reed beds that cover nearly half its surface. On the land around there are tussocks of the great saw sedge and other sedges that give way to willow carr beyond. In this enormous area the plant community seems to have relatively little variety, but it supports up to 280 species of bird, including some that are now rare in Europe. Studies on individual birds and their feeding and nesting habits have shown that there are marked spatial, vertical and seasonal variations in their distribution in the reed swamp. For example, the gizzard of the bearded tit is adapted to feeding on the tough reed seeds in the winter, when it also eats grit to aid the grinding of the seeds. In the summer, when the gizzard is less strong, the bearded tit feeds on water beetles and other water fauna. If at this stage the water freezes over, the bearded tit is unable to feed on the seeds and much of the population starves. Other marked preferences for food or situation were found: the great reed warbler occurred where the reed stems were of a certain diameter, and the great white egret preferred to nest in old and undisturbed reed beds. Herons also nest and feed here, stalking the edge of the reedswamp, or of the pools where the muskrats have temporarily destroyed the reed cover. Most impressive of all is the number of top carnivores – the birds of prey – that this area can support; yet only one of these, the marsh harrier, actually breeds in the reedswamp. Others use the lake and much larger environs as part of their hunting territory.

In this area reed is still regularly taken for thatching, and parts of the lake are used for recreation such as bathing and boating. At present the area is large enough for the wildlife to find alternative sites and the population

numbers have not yet fallen due to human interference.

Distribution of rare species

Many species that are found at Neusiedlersee are distributed in other European communities, although some are rare. The spoonbill, for example, is now found only at Neusiedlersee, Coto Doñana in the south of Spain and on the Rhine in the Netherlands. The small numbers of individuals occurring on the edge of their species' range, such as the swallowtail butterfly in Britain, often get a special scarcity conservation value. Although this magnificent butterfly is common on the European mainland in a variety of habitats, and has caterpillars that feed on several different species of umbelliferous plants such as wild carrot and fennel, in Britain it is only found in fenland, on the Norfolk Broads, and in Cambridgeshire, at Wicken Fen. Here the caterpillar feeds on milk parsley, which is itself an uncommon plant, and although the plant and the habitat have a wider distribution, the butterfly has managed to survive only in these eastern areas of the country. The milk parsley grows among the sedge and reed beds of the fenland, but does not tolerate shading by fen carr trees. While the sedge and reed beds are managed and regularly cut for thatching, the parsley and the butterfly can survive along with many rare wetland plants.

Changing conditions

The Norfolk Broads present a special problem to conservationists which is common in other lowland regions of Europe, such as Holland. The slow-flowing water is highly eutrophic and has accumulated in basins that are now thought to have been medieval peat cuttings. As settlements have increased in this region, so has the amount of waste that is carried by the rivers into the Broads, where it accumulates, creating eutrophic water which results in explosions of the algal population known as 'blooms' (pp. 40–1), which cut out light and oxygen from the lower water. The submerged plants and thousands of fish die, and the floating plants are cut to pieces by the holiday boats that abound in the summer flowering season.

Any conservation management in this region must isolate the reserve from all these problems, as well as maintain a high enough water level for the wetland species. Unusual conditions may in any case occur to destroy many of the plants and animals of a small area (pp. 46–7). If a species occurs in several locations within the area, colonization of the disaster area can occur quite rapidly once conditions have returned to normal, but if rare species are eliminated from a habitat they are

unlikely to return. Thus the size and distribution of populations must be considered when deciding the ideal area of a reserve.

There are many damp meadow species that have been almost eliminated from our flora and fauna in the last century. With the advent of farm machinery, not only could wetlands be drained and improved, but they could also be harvested earlier for hay. The young corncrakes and black-tailed godwits, along with orchids such as the green-winged orchid, went under the mowing machine. Today the black-tailed godwit is a rare bird except in the far northern hayfields of Iceland and Europe where hay-cutting is late enough to allow it to rear its young.

Mountain bogs and moorlands may be sufficiently impoverished to escape agricultural improvement but there are, none the less, other hazards for the wetland species here. Moorlands are frequently managed as grazing for sheep and are regularly burnt to improve the grazing. If this takes place in the winter, little damage to breeding birds or flowering plants is done; but in summer everything, including the dried-out peaty soil, may burn. Man has also attempted to increase production on many of our upland moors by planting conifers that shade out the natural moorland species. This is to provide more useful products for man, so that anyone wishing to prevent this process has to know exactly what he wants to conserve and how to set about it. For any unusual species of the region, he must know the size of its population, its feeding habits and individual territory, and its predators, in order to decide on the size of reserve that is necessary for its survival. This sort of information is being acquired only slowly, but there are now many national and international bodies which are concerned with obtaining and storing the information that is necessary to argue against man's expanding needs.

At long last, man has recognized that wetlands are not wastelands. During the 1960s and 1970s, several positive steps were taken towards their conservation. Much work has been done through the International Biological Programme on wetland sites in Europe. Data have been collected and collated, and special emphasis has been placed on management practices. When this work has been published it will form a yardstick for conservation bodies. In 1971 at Ramsar, in Iran, twenty-seven countries became parties to the Convention of Wetlands of International Importance especially as a Waterfowl Habitat (known as the Ramsar Convention). These countries have listed more than two hundred wetlands deserving special protection.

CARNIVOROUS PLANTS

These strange, insect-eating plants are becoming increasingly rare: do not damage them or collect them from their natural habitats. The easiest place to see them is in botanical gardens.

We are accustomed to the idea that plants are the primary producers of food for the animal world to consume. But carnivorous plants have adopted some very surprising habits, which enable them to reverse the usual relationship between the plant and animal worlds. In order to do this they have sacrificed some of their sun-trapping and food-producing leaf area to specialized mechanisms for catching and digesting insects.

Why feed on insects? The habitat gives us a clue. These plants all live in wet, open places, especially in bogs and heaths, where there is plenty of sunlight but where mineral resources are at a premium because of the very slow decomposition of organic matter in acid soil. Nitrates and phosphates are required by all growing cells: in a mineral-poor habitat, any plant that can collect and break down organic matter has an advantage over its neighbours. In the nineteenth century Charles Darwin cultivated some of these plants and discovered that those which were fed with insects grew much better than those which were not.

An essential stage in the trapping of an insect by a plant is the attraction of the insect to the capture site. Most normal flowering plants attract insects for pollination by their brightly coloured flowers, nectar and nectar guides, or scent. Carnivorous plants use similar devices to attract insects for food. These are formed not from flowers but from modified leaves; their effective life is thus much longer than the flowering period. The golden rosette of the butterwort, the stamen-like tentacles of the sundew leaf and the nectar guides of the pitcher plant, for example, all serve the purpose of attracting prey. Then, once the insect has made the mistake of landing on the leaf, it must be held until death and digestion have taken place. The capture mechanisms used are of two types. There are the passive mechanisms, involving the secretion of sticky and digestive fluids; and the active mechanisms, where the plants have an external 'trigger' device which springs a trap.

▼Round-leaved sundew *Drosera rotundifolia*, a plant of nutrient-poor bogs, has a rosette of attractive leaves covered with sticky tentacles which imprison the insects that land on them.

▼Bladderwort *Utricularia vulgaris* captures small aquatic organisms which trigger off the hairs, so that the bladder opens and sucks in the organism and the water.

▶Great butterwort *Pinguicula grandiflora*, a rarer cousin of the common butterwort, has a pale, shining 'fly-paper' leaf that inrolls on the prey to make a 'temporary stomach'.

▶The vase-like leaves of the pitcher plant *Sarracenia purpurea* have nectar guides that lead the unwary insect to the slippery edge where downward-pointing scales accelerate its fall and prevent it climbing out. The scales are magnified about 250 times.

▼Venus' fly trap *Dionea muscipula* is a plant from the USA. It has elaborate trigger hairs on the inside of the leaf which, when touched by an insect, collapse; the leaf then snaps shut on the victim. The base of the trigger hair is magnified about 120 times.

We expect digestion to take place in the guts of animals where organic particles and tissues are broken down to simpler (and smaller) units in a fluid containing digestive enzymes. The resulting molecules can be absorbed into the body tissues and used by the animal. Perhaps one of the most remarkable things about the plants on this page is the way in which they have imitated this process. The butterwort even forms a temporary 'stomach'! Special glands secrete the enzymes that digest the soft parts of the insect, and then reabsorb the nutrient particles into the plant.

Passive mechanisms

The golden-yellow leaves of the butterwort have a fly-paper surface from which small insects cannot escape; the leaf margins roll over to form a temporary stomach where the digestive process occurs. There are two types of gland on the surface of the leaf: the long-stalked ones, which secrete the fly-paper mucilage; and the hemispherical domes, which secrete the digestive fluids and reabsorb the digested products. The attractive rosette of red-tentacled leaves of the sundew is capable of retaining quite large insects. Once the insect has alighted on the leaf surface, the sticky drops on the tentacles touched retain it, and the surrounding tentacles move towards and imprison it. These glands also secrete the digestive fluids and reabsorb the products. When the tentacles straighten up again all that remains of the insect is its indigestible exoskeleton.

In the case of the pitcher plant digestion takes a long time and may depend on bacteria living in a suitable micro-environment provided by the plant, like the digestive process in our own intestines. The ants or other ground-dwelling insects are attracted by the nectar secreted along guide-lines up the side of the pitcher-shaped leaves to the slippery edge of the pitfall trap. The pool of digestive fluid into which the insect falls may already contain other insects. One leaf may catch many insects, and the solution becomes a food resource from which minerals are withdrawn when required.

Active mechanisms

The exotic venus' fly trap comes from the coastal plains of North and South Carolina in the USA, but is now generally available as a cultivated plant. Hopping and crawling insects or even a paintbrush can trigger the hinge mechanism between the two sides of the leaf. When the hairs on the leaf surface are touched, the flexible coupling gives way and the leaf snaps shut with the bristle-like teeth on the margins overlapping across the opening. If the leaf has caught organic food material, glands on the surface of the leaf secrete a pool of digestive fluid.

Bladderworts generally grow in nutrient-poor freshwater ditches and moorland pools, where the water supports a great variety of water fleas and other small organisms. The compound leaves are formed of finely divided threads, and some of these are modified into bladder-like flasks with a specialized trap mechanism: water is actively pumped out of the bladder with the trap-door shut, creating a negative pressure; when a small animal touches the trigger hairs at the entrance, the door springs open and the creature is sucked in.

1 Venus' fly trap plants are available at many garden centres. Obtain one or more plants and set up experiments to test which substances stimulate the secretion of digestive fluids and what happens subsequently. Try 'feeding' the leaves with small insects, bird droppings, fragments of cheese and drops of liquid proteins such as raw white of egg and milk. Do the enzymes cause egg-white to coagulate or milk to curdle? If you put any of these substances on the leaf very carefully, so as not to spring the trap, will an open leaf secrete digestive fluids? Some of the substances will change anyway in the course of time; as a control to check this, put similar samples on a non-secreting leaf, such as ivy, or even on a plastic model leaf for comparison.

2 Take a small area, about 2 metres square, in a boggy place where sundews grow and look at the insects that are trapped on the leaves. Count the number of leaves on each plant and the insects caught, note the types of insects and record the plants as shown.

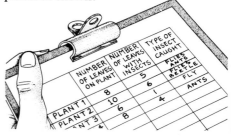

Sort the insects into the following groups. Sometimes it is easier to identify small insects if you put the leaf into a tube of methylated spirits to kill the insects before looking at them with a hand lens.

Flies and midges: insects with 2 wings and piercing or sponge-like mopping-up mouthparts.

Craneflies and wintergnats: larger insects with 2 wings and long legs.

Beetles: insects with 4 wings, 2 of which are hard wing cases protecting the flying wings.

Moths or butterflies: insects with 4 conspicuous wings and long antennae.

Ants: insects without wings (except for a short period in the summer) and body with obvious head, thorax and abdomen.

Springtails: small, wingless, litter-living insects that spring across the surface.

Spiders: 8-legged members of the arthropods (all insects have 6 legs) with head and thorax fused together.

If you come back at another season or in different weather conditions, are your plants catching different insects?

MORE WATER

The amount of water used per person each day goes up annually. Where does it all come from, and what effect does our management of it have on freshwater communities?

Man uses water far in excess of his own biological requirements. An automatic washing-machine uses around 65 litres a wash, compared with 15 litres used in hand washing, and as more and more families acquire washing machines so the domestic consumption goes up. Water is required in industry for three main purposes: as a solvent; for cooling, as in power stations; and for hydroelectric power. In 1956 overall water consumption in Britain was increasing at the rate of 2.3 per cent per annum, but by the early 1980s it had risen to 8 per cent per annum. By the year 2000 a city with a population of around half a million people will have doubled its present consumption. Where can all this water come from?

Water arriving on the surface of the earth as rain is almost pure, but accumulates impurities and organisms on its journey to the sea. It is therefore cleanest near the source, so many reservoirs are sited in upland regions where rainfall is high. These reservoirs are often situated far from the centres of population that they serve, so the water has often to be piped for considerable distances. Water can be collected at any point on its journey to the sea, but in densely populated lowland regions, the water collected must be treated to remove the undesirable living organisms as well as the chemical impurities that it contains. Water can also be extracted from the rivers themselves, but as this supply varies throughout the year, careful regulation of any river supply is needed. It has been suggested that connections between rivers would overcome this problem, but this would result in the mixing of water and organisms from different rivers and would have a devastating effect on plant and animal populations that are dependent on a particular pH or nutrient supply. Finally, water may be extracted from the water table underground, where this is abundant, but care must be taken that this neither interferes with the supplies from surface springs, nor lowers the water table so much that crops and natural vegetation suffer drought.

▶An upland reservoir in the Elan valley in mid Wales. The mountain streams that feed this reservoir carry sediment but few nutrients from the moorland around them, providing an oligotrophic lake of 'clean' water.

▼Grafham Water, a lowland reservoir in Bedfordshire, England, surrounded by farmland, provides a nutrient-rich water supply for many species. This water requires treatment, after which it is pumped up to the tower in the background to give it a 'head'. Many lowland reservoirs also add amenities such as fishing and sailing to the area.

▼In the densely populated Mediterranean peninsula of Gibraltar, water is a precious commodity. This extensive water catchment area has been built so that all the rain water can be channelled into huge subterranean reservoirs in the heart of the limestone rock.

▲Taf Fechan reservoir in south Wales in August 1976, when drought conditions reduced the input to this reservoir to one small stream. Previous fluctuations in water level are obvious on the shingle shore, as well as a fringe of weedy plant colonizers. Many reservoirs are afforested around their shores to reduce sedimentation and nutrient input from the surrounding land.

Cleaning and recycling

Artificial water bodies such as reservoirs, like any new water body (pp. 36–7), soon accumulate wildlife. Any organisms present that would be detrimental to the consumer must be removed before the water is pumped into the mains water supply.

Most organisms and sediment can be removed by settling and filtration processes, which occur naturally as the water passes through the soil, but where there are many microscopic algae in the water the removal of these may be very costly as the algae continually clog the filters. In many major cities water is collected and recycled several times before it is discharged, but this involves removal of chemicals such as phosphates, which is done by an expensive chemical stripping process.

The effect on wildlife

In many ways the collection of water in artificial lakes or reservoirs greatly increases the habitats available for larger organisms such as birds, particularly in the lowland areas around cities. In London, ornithologists have recorded many new birds since the construction of the major reservoirs at the turn of the century. In 1902 a naturalist visited Barn Elms, an inner London reservoir, four years after its construction, and was delighted to see black-headed gulls, tufted ducks, great crested grebes and coots. Today a naturalist would be disappointed to see fewer than ten species, including one or two unusual regulars such as goosanders and smews. Birds take time to get accustomed to the new sites that offer them good places to breed or stopping-off places on long migration flights. Some birds, like the tufted duck, have greatly increased their breeding range and number due to the ever increasing number of habitats available.

Although birds can exploit the water in a reservoir, there are many animals and plants that cannot survive the extreme conditions imposed by management. The water levels can vary with the dam height, the catchment area and, most important of all, with the amount of water that is withdrawn at any one time – the drawdown capacity. In summer, when less rain falls, upland reservoirs may be almost dry when their water is transferred to systems lower down and nearer centres of population. Thus a reservoir is an unstable habitat, offering a large space to colonizers and those that can survive the occasional drought conditions (pp. 46–7). In lowland reservoirs where abundant nutrients are available in the run-off water from the surrounding farmland, there are liable to be explosions of algal and planktonic populations, which kill other species through shading and deoxygenation of the water. They may also impart a bad taste to the water and cause problems in filtration.

There is rarely a fringe of emergent plants around the edge of a reservoir, as these require some permanent water, at least over their roots. After a large drawdown, a reservoir can often resemble a weedy garden, its old shore lines covered with terrestrial plants, many being aerial colonizers, such as the willowherbs. Other plants, such as the brooklime and yellow cress, may be brought in by the stream which feeds the reservoir and become abundant for a period.

The stream below the reservoir is also affected by the drawdowns of water as they create a sudden flood situation that may wash out much of the invertebrate population as well as changing the nature of the stream bed. If this occurs in the breeding season, it may take many years for the stream community to recover. The animals may also receive a temperature shock as the water is released from the colder, deeper layers of the reservoir (pp. 16–17).

For man himself the reservoirs have added many amenities to the countryside, such as sailing, which many water authorities allow, and fishing, where they are kept stocked with trout. The fish also control the invertebrate populations (pp. 42–3). But the purpose of a reservoir is to provide drinkable water, so these amenities must in no way lower the quality of the water.

① Selecting sites for reservoirs needs much background knowledge. The bedrock must not be porous, and must support the dam. The catchment area must be large enough, the rainfall high enough and the valley deep enough to hold a large amount of water when it is dammed. Then there are the problems of land to be acquired and the management of the land adjacent to the water. Many water authorities allow public access to reservoirs today, and also produce information on the capacity, use, etc. of their reservoirs. Visit one of these reservoirs with this information and map out roughly the wildlife habitats that are available around the edge of the water (meadow, scrub, deciduous woodland or forestry), and the shallow and deep areas of water. Visit it regularly to see which birds use the open water and the edges. How many are migrants/residents? Where do they prefer to feed? Are they associated with a particular habitat?

Many water birds frequent reservoirs, both as visitors and as breeding birds. The construction of new reservoirs has rapidly extended the breeding and overwintering range of the tufted duck *Aythya fuligula*. It feeds in the shallower waters, diving for molluscs and invertebrates, and nests at the water's edge near scrub or reedy pasture.

② Sample a stream above the reservoir and below the dam for invertebrates using the kick sampling with a net (pp. 14–15) and the stone scrubbing (p. 73) methods. Which organisms are missing below the dam? If you can visit the stream below the dam after a drawdown of water, repeat the sampling. Is there any difference? (If you are able to do this you could also sample the stream with a core sampler as on p. 85, and count the organisms, to find out the proportions of populations that survive.)

ANATOMY OF A RIVER

The change from a clear, rushing mountain stream to a cloudy, sluggish river, meandering through its own deposits to the sea, is accompanied by changes in the life that inhabits the river.

Rivers, which are ancient features of our landscape, take many millennia to cut their valleys and build up their plains, but today their valleys and plains are managed by man up to the banks themselves. Britain's longest river, the Severn, is only about 350 km long, compared to the 2,832 km of the Danube, the longest European river. Yet, on its journey from the Welsh mountains to the estuary, the Severn travels through the typical stages of most rivers, producing habitats that are associated with particular organisms and fish. A good angler will tell you precisely which fish he expects to catch at each stage of the river's journey.

The Severn rises at around 600 metres on the east slopes of Plynlimmon. The waters of many acid upland streams merge in the steep, V-shaped valley between the hills and the flood plains below. The potential cutting action of the stream depends on the amount of water and the gradient of the slope, which in turn depends on the rainfall and catchment area of the stream and on the hardness of the bedrock on which it is working. Rainfall may vary from season to season and from year to year, as well as over the centuries. In temperate Europe the highest rainfall usually coincides with the winter or cold season, so the Severn generally has most water in it in the winter and least in the summer. The ammunition used by the water is the rock itself; boulders and pebbles are hurled at the bedrock, the erosive power of the water increasing with the size of the material in the stream. This sounds a potentially dangerous habitat. Yet the erosion is not even. Deep pools form where the water seems to pause for a while in its rapid descent, and here you may surprise a small trout, darting under the banks as you appear.

These waters are well endowed with oxygen, but there is little obvious food for the inhabitants as all organic matter, apart from a few mosses attached to the more stable boulders or to the bedrock, is carried downstream. The oaks that used to contribute their autumn leaf fall

◄Once in the foothills the Severn begins to deposit its load of shingle, especially where the river Dulas adds its load to the river. Most of the shingle is regularly disturbed by storm flow from the hills, and is consequently bare of vegetation.

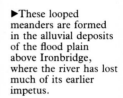

►Sun glints on an old course of the river, now isolated from the main stream. Old river terraces are left behind as steep banks when the river changes course.

►These looped meanders are formed in the alluvial deposits of the flood plain above Ironbridge, where the river has lost much of its earlier impetus.

◄The river winds through the wider valley, constantly altering its route through its own deposits, and leaving traces of these changes.

◄The river has gathered many tributaries and dropped most of the shingle. It flows wide and deep across the flood plain, cutting steep banks around 3 metres high, even on the pasture side. Where the banks are higher, on the outside of the bends, scrub and woodland develop.

to the stream have been replaced by a forest of spruce whose tough, acid needles cannot contribute many nutrients. Yet, if you turn over some of the water-worn and polished rounded stones, or look among the strands of moss, you will find some small inhabitants keeping out of the current's reach (p. 72). For thousands of years the salmon have come up here to spawn in the cold, oxygenated, upland waters. They are still able to find their way up a river that has been so much altered by man in respect of gradient, water content and chemistry, and even manage to negotiate man-made obstacles such as weirs.

As soon as the gradient falls, the river begins to drop its load, first of shingle,

(riffles) followed by slower, deeper areas with a finer sediment on the bottom (flats). In early summer the riffle areas are picked out by the flowing green tresses of water crowfoot, covered with white flowers above the surface. These plants prefer the oxygenated water here. Other organisms, including fish, are associated with either the riffle or the flat areas. Trout are still found in the riffle areas but grayling and minnow are the characteristic fishes of the flat areas, and, as the pH of the river increases here, bullheads and loaches appear as well. As the gradient falls, the shingle banks and spits, dumped in storm flow, are less frequently disturbed and have become covered in a rich, herbaceous vegetation.

The river gathers more water from tributaries and moves into its flood plain, where the landscape and river pattern change dramatically. This region of rich alluvial earth formerly supported great oaks, but today field after field produces quantities of grain and other arable crops. This land used to be flooded regularly by the river, but since the creation of reservoirs in the uplands the flow of water is now regulated and floods are rare today. The river winds and curves through this landscape, increasing the curves by eroding the bank on the outer edge and depositing material on the shallower inner edge, until the water breaks through the banks taking the shortest route and leaving the old curve as a stagnant pool, known as an oxbow lake. The river incises deeply into the landscape running between steep, high banks, that are often wooded, and yet everywhere there is evidence of other routes, as in the old river terraces that are now banks or field boundaries. The deep, murky water of the now extensive flats is nutrient-rich and deficient in oxygen, but the silty bottom supports many varieties of pondweeds, water-milfoil and other water plants. Near the top of the food chain are many coarse fish, such as chubb, roach, dace, perch and pike, and now also the barbel, which is more typical of European mainland rivers, but was introduced in the lower Severn in 1956 and has been increasing and spreading ever since. Another continental fish that has been introduced recently is the zander. A native of the Danube and Rhine, it is equally at home in the muddier reaches of the Severn.

In the murky, silt-laden water of the lower reaches, fringes of emergent water plants appear; here the bream is the characteristic fish. This water now receives the waste products of many of the lowland settlements, lowering oxygen levels, raising temperatures and increasing the amounts of dissolved chemicals, allowing very few aquatic species to survive.

then of gravel, sand and finally silt. This fall in gradient is associated with an increase in man's agricultural activity. The rolling hills with their patchwork of fields support a sheep and cattle farming community that confines the woods to the steeper slopes. Agricultural fertilizers and chemicals find their way into the river, but the water is also changed by other factors. Tributaries join the river, bringing water from other geological regions. If these contain limestone rocks, the water is of a higher pH and has a corresponding richness of species (pp. 66–7). The material dropped by the river constantly obstructs its flow, producing a pattern along the length of the river of fast, shallow, shingly areas

PROJECTS

1 Make a map of a river that you can visit, including all tributaries and settlements associated with the river. Add as many environmental details as you can, such as contours, geology and land management. Then make field visits to different bridge sites to record width, current speed (p. 15), sediment and bank type of the river, as well as the plants growing in it. If necessary collect these with a grapnel, but beware of trying to sample organisms at the edge of larger rivers, as the banks may be unstable and the current fast. Find out from your local angling societies which fish are caught in different parts of the river. Can you find any association between the river type and the fish in it?

An upland settlement, Llandinam, clusters around the river where the narrow valley confines its route. The shingle deposits are now well vegetated.

2 Try making your own river patterns in a sand pit or tray with a depth of at least 50 cm of gently sloping sand. Use a hose-pipe fitted with a sprinkler to simulate rain over a marked area of the sand (the catchment area). At first the water will soak into the sand (ground water) but as the sand becomes saturated the water will run off, finding its own channels. From where is sand removed, and where is it deposited? Can you change this by increasing or decreasing the amount of water falling? What shape are the valleys in the areas of erosion and deposition?

STREAMS AND BROOKS

Water passing over hard, resistant rock for millennia has gradually cut this V-shaped valley. Although this small stream appears to carry very little water now, a few days of heavy rain will dramatically alter it to a torrent of water capable of carrying some of the boulders visible on the stream bed.

On these bracken- and gorse-covered hills the soil is thin and retains very little water, so the stream will rapidly receive the rain water running off the catchment area around it. This water carries mineral and organic matter from the valley slopes, transporting it downstream to lowlands where the gradient is not so steep. The net result of this activity is to remove the highlands and fill in the lowlands.

However, the route through the valley is not straight, and the stream is forced to make many twists and turns where it meets harder bedrock. In the foreground there is a sill of harder rock damming the stream and causing the water to find its way round it. If the rock below the sill is much softer, a waterfall may eventually form, as has happened at Svartifoss (p. 12).

A highland stream in Ashes Valley on the Long Mynd in Shropshire.

VARIATIONS ON A THEME

The character of each stream or river has been and continues to be shaped by the interaction of water or ice and the landscape.

There are many factors affecting both the shape of the stream bed and its route through the landscape, as well as the potential animal and plant habitats that it provides. Some of these are not obvious at first sight or in one place; a tinkling summer brook may in winter become a raging torrent which carries many organisms and large quantities of sediment far downstream. Storm flows may leave large boulders in or near the stream, and these often remain in place long enough to acquire a mossy cap. But not all boulders arrived in this way; many were deposited under other conditions, such as by the movement of glaciers, not by the stream that makes its way around them now.

Today's variables

The volume of water that a stream carries varies greatly from season to season and from year to year. This is dependent both on the rainfall and on the permeability of the surrounding ground or bedrock; the less permeable the rock, the more water runs off to form streams. Where the ground surface is permeable the streams appear only where there is an impermeable layer beneath. In cool, temperate Europe most of the rain falls in the winter season; in the more northern, colder regions a large amount of the annual precipitation may be locked up as snow until the spring thaw. In contrast, valleys in warmer, drier latitudes may dry up completely in summer, causing harsh extremes for plant and animal life.

The cutting or erosive power of river water is a combined effect of the water volume, the gradient of the river bed, and the sediment load. But steepness is itself associated with the hardness or softness of the rock on which the water works. Thus steep valleys with rushing, boulder-filled streams are associated with upland areas of resistant rock, and slow, meandering streams are usually associated with softer rocks or alluvium of old river or ice deposits.

Erosion in limestone regions gives a very different pattern, however. Here much of the action of the water occurs through solution of the rock, so that erosion takes place downwards rather

▲Spring rain on the uplands has filled this river so that there is little to be seen of the banks of rich alluvial material through which it cuts.

◄An eroding upland stream, running through shady deciduous oak woods in Lancashire, in the north of England, carries rocky material of all shapes and sizes. The largest boulders become covered with mosses in between the occasional storm flows that move them.

◄In early summer this river is full of water crowfoot *Ranunculus* sp and the banks that are not being eroded quickly support a variety of species, including hemlock water dropwort *Oenanthe crocata*, hairy willowherb *Epilobium hirsutum*, and comfrey *Symphytum officinale*.

▶A mountain stream, half hidden by overhanging peaty banks of heather and rushes, is brown with iron and contains plenty of bog pondweed *Potamogeton polygonifolius*, which tolerates the high acidity and frequent changes in water level.

than sideways. This forms not only the typical gorges, such as those of Cheddar in England and the Dordogne in France, but also the rounded swallow holes and their underlying cave systems, which are carved out of solid rock. Chalk, limestone and any calcareous rocks in turn affect the chemistry of the water by raising the pH with the dissolved calcium. This increase in alkalinity is associated with an increase in nutrient and mineral availability, creating a much more desirable habitat for many plants and animals. Much of the water in chalk and limestone areas is filtered during its long journey through bedrock, and emerges purified of many pollutants. Consequently chalk and limestone streams are abundantly rich in plants

▼A slow-flowing river makes its way across lowlands on boulder clay, which gives this water its rich, eutrophic character. Submerged and emerged plant species abound, including many that change their habit during the year, such as arrowhead *Sagittaria sagittifolia*, which produces submerged, then floating and emerged leaves with the flowers.

and animals. Chalk streams in Hampshire in the south of England provide some of the most expensive and famous trout fishing available (pp. 92–3).

Landscapes of yesterday

On finding a peaty stream winding across a wide highland valley strewn with large boulders, or a small stream meandering through a valley rich with permanent pasture right to the stream edge, you may well suspect that the present streams have contributed little to the shape of the landscape around them.

Rainfall and climate have changed dramatically over the last 60,000 years. Where the rainfall was higher the erosion and deposition were greater, so valleys are frequently filled with old river deposits of shingle or sandy alluvium. A stream running through such a valley today erodes the banks at water level; the unstable alluvium above constantly slips into the water, creating a steep, vertical bank where there is no edge available for fringe plants, apart from the trampled shallows made by cattle coming for water.

In many areas, however, the last time that the landscape was greatly altered was during the Ice Ages. The changes were brought about by the carving and carrying action of the ice itself, by the moraines churned out from the ice sheets, and by the fluvioglacial deposits that accompanied the melting of the ice. Parts of Denmark are formed entirely from these deposits, and much of Britain and Northern Europe is covered with deep deposits of impermeable boulder clay, a combination of ice-carried boulders and ice-ground 'rock flour'. Rivers must traverse these areas before reaching the sea; but the water slows down on the wide plains, allowing the growth of a broad fringe of emergent water plants, such as branched bur-reed and the beautiful flowering rush, as well as of those that have submerged and emerged leaves such as water plantain, arrowhead and club-rush. There will also be the floating plants of duckweed and frogbit proliferating in the slow-flowing waters. These rivers are rich in organic nutrients and highly eutrophic, so that only the fauna that can live in oxygen-deficient waters can survive here.

Where ice action has left depressions in the highland regions, nutrient-poor peaty trickles can be found, often rising in boggy areas full of sphagnum and other moorland plants (pp. 62–3); but these runnels of water gather into streams overhung by banks of heather or rushes. The boulder-strewn areas alternate with deep, peaty pools that are full of bog pondweed and floating club-rush, both plants that can withstand the enormous variations in water levels that are associated with highland areas.

1 Ice and frost erosion produces angular pebbles, while water erosion produces rounded ones; but where the river carries ice-deposited material it takes time to wear off the angles. Find a river in an area where ice action is known to have occurred, and collect 50 pebbles in the size range of 5–10 cm, rejecting all split or broken pebbles. Use the chart below to sort the shapes into 4 categories, and plot the percentage in each category on a histogram. Compare it with those below. What type of deposit have you got?

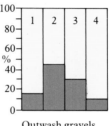

Head
(angular fragments formed by ice and frost)

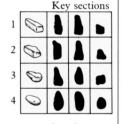

River gravels
(water-worn pebbles)

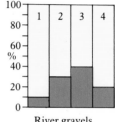

Outwash gravels
(pebbles showing ice and water action)

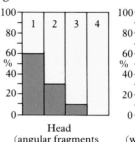

Key sections

1 Angular
2 Sub-angular
3 Rounded
4 Well rounded

2 Make a transect across the curve of a shallow stream where you can measure flow (with Thrupp's nails, as described on p. 15) and depth, and record the type of sediment on the stream bottom (sand, gravel, silt, etc.). Repeat the transect in different weather conditions or seasons. Does the sediment type vary with the current? Where are the largest stones and finest sediment on your stream bed?

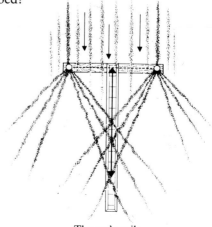

Thrupp's nails

GROWING IN A CURRENT

Living in mid-stream is like taking part in a continual tug-of-war with the downstream current. Often, with the extra water of a storm flow to help, the stream wins.

Water-milfoil *Myriophyllum alterniflorum*, with finely divided leaves

Water starwort *Callistriche hamulata* with strap-like submerged leaves and rosettes of oval floating leaves

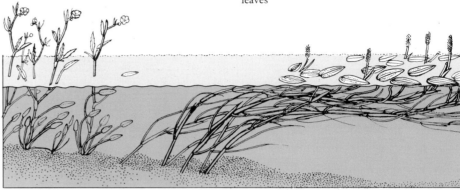

Lesser spearwort *Ranunculus flammula* will produce a diversity of leaf shapes according to whether it grows in deep or shallow, swift or slow water.

Broad-leaved pondweed *Potamogeton natans* has thread-like submerged leaves and floating leaves with a 'joint', or bend, in the stalk, just below the blade, that keeps it floating flat on the surface.

Nothing stays the same for long in this environment—the stream bed is always being either removed by erosion or covered up with sediment. The speed, amount and clarity of the water can vary from season to season or from day to day.

Plants that live in this 'difficult' habitat need to be versatile and tenacious. Yet you will find growing here the most delicate and fragile of plants, which it seems almost impossible to imagine surviving the constant battering of the water. Their secret lies in offering the least resistance to it. Large wading boots offer a broad, solid front to the current which directs its force against them, but the delicate, flaccid water plants take on the shape that the water dictates, and are streamlined and supported by it. Their leaves are reduced to thread-like or strap-like parts, and a clump of these will slow down the current so that sediment carried in the water drops into, or downstream of, the clump. The stalks often break easily, an adaptation to a sudden increase in water velocity whereby the shoot is lost but the root remains to grow again. If the clump gets too large and solid the whole plant is likely to be lost. The water carries most of the nutrients that an aquatic plant requires, but underground stems and roots are essential for maintaining a plant's situation in the current. Those that grow mainly in the surface sediment are likely to be washed away in a storm, but those that have deep underground stems or roots, like the unbranched bur reed, can continue to put up new shoots after storms have damaged the old shoots. They can also withstand burial by a sudden increase in sedimentation.

The rate of flow varies considerably across a stream, as do the conditions from stream bed to surface, and the growing efficiency of a plant will vary according to its position in the water and its leaf shape. Different types of leaf are efficient in a submerged situation, a floating one and an emergent situation. Thread-like

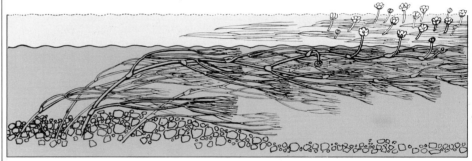

Water crowfoot *Ranunculus fluitans* with long, thread-like leaf segments that move with the current. The flowers are aerial but the fruiting stalk bends back into the water.

▶ A clear chalk stream with clumps of water starwort *Callitriche* sp and water crowfoot

Ranunculus sp streaming in the direction of flow.

submerged leaves become tangled at the water surface, where an oval boat shape is more efficient both in use of light for photosynthesis and in resistance to the surface disturbance, especially if it has a waxy, unwettable surface (p.30). There are very few emergents that grow in fast streams, but those that survive have tough, sword-like stems or leaves that offer little resistance to changes in flow or water level. Some plants produce different leaves in each situation. One of the commonest water plants, the lesser spearwort, produces strap-like, submerged leaves, oval, floating leaves and spear-shaped, toothed, emergent leaves, all at the same time! The broad-leaved pondweed produces linear, strap-like, submerged leaves and oval, floating leaves with a 'joint' at the top of the stalk which maintains the leaf blade at the right angle to the surface when the water level changes. Many water plants produce the floating leaves only

in the summer season when there is little risk of surface damage, and maximum benefit from the surface sunlight. The water crowfoot produces broad-bladed, floating summer leaves which also, like a raft, support the numerous white flowers in early summer.

Reproduction problems

Pollination This is one function that few water plants have managed to make aquatic. Pollen is easily damaged by water, so pollination is still generally an aerial affair. But adaptations to wind and water dispersal are very similar: for example, petals get in the way, so are reduced. Water starwort has managed to become a true aquatic by producing pollen containing water-resistant oil in a single stamen on male flowers near the surface. This drifts down on to the long styles of the females that are in separate flowers lower down the stem. If you

1 Test the breaking strength of the leaf stalks of different submerged plants by attaching a spring balance to the whole leaf with masking tape and thread. Do not use thread alone as this would create a stress point.

2 With waterproof tape, try attaching different shapes and surface areas to a stalk, and see the effects of running water, in either a stream or a model. (Remove the growing leaf first.)

3 Water crowfoots present problems of identification because of their apparent similarity. Choose plants that grow near you, make habitat notes on the stream, velocity, depth, width etc., and plot the variation in characters, such as leaf/stalk ratio, leaf length/internode length ratio, number of divisions in each leaf, shape of leaf in the water. Does the distribution of characters vary with the growing conditions of the plants? Can you identify the species with the help of one of the recommended books (see p.188)? If you have two or more species do their leaf characters overlap? (Pondweeds are also suitable material for this project using characters like leaf length/breadth ratio, stalk length, venation patterns, stipule length and shape.)

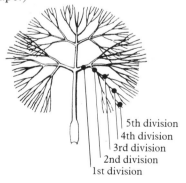

5th division
4th division
3rd division
2nd division
1st division

leaf length

internode stalk

leaf blade

take out a piece of water starwort in the summer you will find all the lower leaf axils contain fruit. Water crowfoots are successful aerial reproducers whose sheets of white flowers cover streams and rivers in June. This simple white flower that produces nectar and abundant pollen attracts insects that are hatching everywhere at this time and when fertilization is effected the fruiting head bends back into the water to release the achenes, or seeds. But it only takes a sudden change in water level or a rough surface to drown these white carpets.

Wind pollination is quite common in certain groups, such as the pondweeds, some of which produce occasional spikes of petal-less flowers above the surface of the water to be pollinated by the wind, and others of which are submerged and pollinated by water. In both these groups there are often very few seeds set. So it is hardly surprising

that many of our water plants reproduce vegetatively either by a rapidly expanding underground stem or from small fragments that break off and drift downstream.

Dispersal If pollination of water plants is usually aerial, fruit dispersal is usually aquatic, the only unusual problem being in preventing germination in water until dispersal has taken place. Many seeds have tough, water-resistant coats and may float for several days or longer before sinking to a suitable place (p.82). Sedge seeds are a favourite food of water birds, and pass through their digestive systems; other seeds are carried to new places on feet or feathers. But birds also carry small portions of water plants to other places which then spread vegetatively. Canadian pondweed has spread rapidly in Europe, although only the female plant was introduced (pp.98–9).

69

SWIMMERS AND CLINGERS

Many animals behave in such a way that they avoid the perpetual tug-of-war against the current. Others have become adapted to living on the tops and sides of stones, apparently in the full force of the water.

Hydrodynamics

Careful measurements of current velocity, taken from the surface of the stream through to its bed, reveal that the rate of flow decreases towards the bottom. On the surface of the stones is a narrow layer, 0.5 to 1 mm high, where the water is static, or, at most, moving very slowly indeed. This is known as the 'boundary layer', and many of the animals which live on the tops and sides of stones have adaptations enabling them to exploit this microhabitat.

Typically, these animals have very flattened bodies, as, for example, the flatworms and leeches. The flatworms secrete a slime track from the undersides of their bodies, into which beat many thousands of microscopic, hair-like processes, or cilia. Their wave-like action draws the animals in an apparently effortless gliding motion across the substratum, and they also act as miniature anchors, holding the animal in place. This attachment mechanism is so effective that many pond-dwelling flatworms may be observed gliding upside down beneath the surface film. The flattened leech body has a pair of suckers (one at each end) which holds the animal in place.

Clinging on tight

Suckers are also used by a number of other organisms. The blackfly larva *Simulium* occupies the tops of stones in those parts of the stream where the current flow is quite fast, and must be regarded as one of the most successful organisms at maintaining its position in the full force of the current. It may be said to belong to the belt-and-braces brigade, for it utilizes no fewer than three ways of attaching itself. There is a sucker at the base of the body and a smaller one on the end of a short, fleshy proleg just behind the head. Inside the walls of the posterior sucker there are between 4 and 64 horny hooks.

A *Simulium* larva attaches itself to a

▼A stonefly nymph *Perla bipunctata* at rest under a stone, showing the feathery yellow gills and the pale fringes of hair on each leg, which aid rapid movement.

▶The water snake *Natrix maura* swims through the water in Portugal with a sinuous movement of its body.

▶A water mite *Arrenurus* sp swims under water with few adaptations to swimming or clinging other than its small body size.

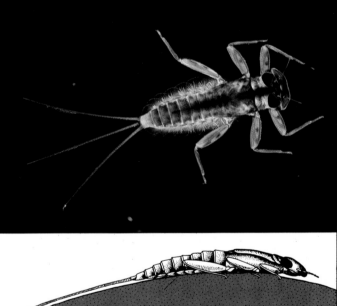

◀The broad, flattened head of this mayfly nymph *Rhithrogena semicolorata* faces into the current, so that when the insect is at rest on a stone it offers least resistance to the current. It also has the first pair of gills modified into a sucker to maintain its place, even in the strongest current.

▼These blackfly larvae *Simulium erythrocephalum* are attached to a water crowfoot leaf segment by a sucker on their hind end that is lined with many chitinous hooks. The proleg just behind the head also ends in a sucker, enabling the larvae to live and move around in the fastest current without getting washed away.

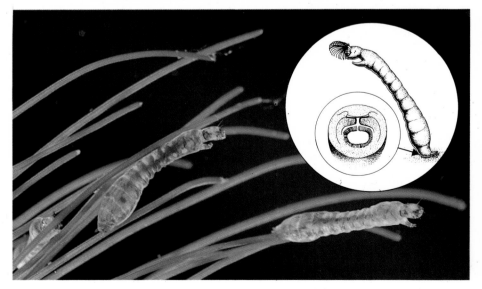

suitable rock by using the larger, posterior sucker; in a fast current, the smaller, anterior one is used as well. Using a secretion from the salivary glands, the larva then spins a small, silken mat over the surface of the stones by throwing its head backwards and forwards. Once the mat has hardened, the larva stretches over it, attaching itself by the anterior sucker, pulling the posterior one over the mat, and attaching it. The hooks within its wall then latch into the silken threads cemented to the stone's surface. The anterior sucker is released and the animal stands up in the current. A silken thread is often attached from the mouth to the substratum so that if the animal should be swept away it can pull itself back along this anchor rope, using its proleg, and re-attach itself.

A sucker is also utilized by one of the clinging mayflies *Rhithrogena*. In this case it is formed by the modification of the first pair of gills which are tucked beneath the body to form the walls of a simple, sucker-like device. Not all clinging mayflies possess such a structure but all are beautifully modified for life on the tops and sides of stones. Their bodies are hydrodynamically shaped: the head, which is kept constantly facing into the current (positive rheotaxy) is broad and flattened, as is much of the rest of the body. The animal is tallest about one-third of the way along, then tapers back rapidly down to the tail. As water flows over such a body it tends to force it down on to the substratum. The pattern of water flow closely resembles the slipstream of air over a racing car. The legs articulate out to the sides of the body rather than beneath it, again ensuring that much of the body will lie within the boundary layer, and they terminate in long, powerful claws. (This pattern of leg articulation is also seen in the small, black riffle beetles). The mayfly's three widely spreading 'tails' are covered with pressure receptors, and the information they feed back to the brain helps to keep the animal facing directly into the current.

Some of the animals living on the tops and sides of stones seem to show relatively few adaptations in shape. The caddis larvae, for example, inhabit large, bulky cases (pp. 76–7) which project through the boundary layer and into the faster-flowing water above. The animals cling on tightly with their strong claws, but it is the density of the case which holds it in position. It is particularly noticeable that caddis larvae with cases made of plant material or small sand grains are generally more common in slow flowing sections, while the larger, stone-cased caddis occur more abundantly in faster flowing water. In very fast flows, they may anchor the case down with silken thread.

The strong, muscular foot of the river limpet anchors it to rocks in a fast flowing river. When individual limpets from fast flowing streams and from slow ones are compared, the heights and breadths of the shells clearly differ. Another mollusc found in larger rivers is the zebra mussel, which anchors itself down using countless fine, byssal threads, just like the mussels found on the sea shore.

One group of animals which show little or no adaptation in body shape is the water mites, whose globular bodies are so small that they lie completely within the boundary layer.

Swimming through the water

The ability to move rapidly through a zone of fast-flowing water is often an essential part of survival for many of the predators living in the stream and for those animals which have to commute from the stream bed to the surface in order to breathe. Through time, many of these animals have evolved a rather similar hydrodynamically shaped body which minimizes the resistance to the water flowing over it. As in the clinging mayflies, the body is broadest about a third of the way behind the head and then tapers steadily towards the back end. The body may then be propelled forward by sideways sinusoidal movements, as by fish (trout and eel) and snakes (water snake), or by vertical movements as by leeches, otters and the free-swimming mayflies such as *Baetis*; or by the rapid movement of legs which are frequently covered by hairs, as in the case of many water beetles and water boatmen. Large dragonfly nymphs propel themselves rapidly through water by forcibly ejecting water via the anus (pp. 24–5).

PROJECTS

1 There is a suggestion inherent in this and the following spread that different animals occupy different areas of the stream bed. Using a pond net, prod about vigorously for 30 seconds on a stony section where the current flow is rapid. Repeat in a silty area. Compare the animals found on each type of bed. Repeat several times. What animals do you capture if you sink two terylene doormats, one on a rocky bottom, one on a silty one, for 4 or 5 days; or if you secure strips of polythene 5 cm wide and 50 cm long at one end in a fast-flowing section and a slow-flowing one?

2 Swimming close to a neighbour brings about shoaling in some fish (pp. 164–5). Take one stickleback and place it in a clean kitchen sink filled with water. Place a stick across the sink so that it divides the sink in half. Observe one half for 5 minutes, and, using a stop-watch, record how long the fish spends in that half. Repeat 5 times. Work out how long the fish spent in each half of the sink. Now suspend a second stickleback in a small polythene bag full of water at one end of the sink. Over a 5-minute period, time how long the original fish now spends in this half of the sink. Repeat 5 times. Is there any evidence that the fish spends more time in one half than the other? What happens when you add two or three more fish to the polythene bag?

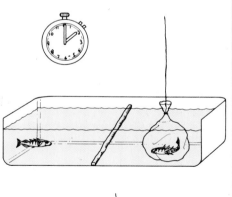

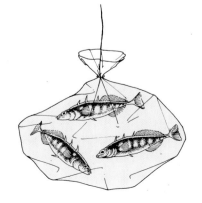

BURROWERS AND HIDERS

Some animals living in streams
and rivers avoid being swept
away in the current by burrowing
into the soft bottom sediments,
or by hiding beneath stones or
among clumps of weed.

◄A burrowing mayfly nymph *Ephemera vulgata* showing its tools for the job: two tusk-shaped mandibles, and the flattened, hairy forelimbs that dig out the burrow. The branched gills are curved over the nymph's back.

▶The freshwater shrimp *Gammarus* sp is one of the most abundant freshwater invertebrates in a stream. Although it is flattened from side to side, it has no way of maintaining its position in a current except by hiding below stones or among weeds.

◄The bullhead *Cottus gobio* is camouflaged near the pebble stream bed, where it waits to rush up and seize prey passing above it.

▶The predatory dragonfly nymph *Cordulegaster boltoni* waits on a muddy bed for unsuspecting prey to come near, for the hairs on its body trap silt so that the nymph is concealed. The 'mask' with which it catches its prey is concealed below the head in the photograph.

▶One method of staying in position among the submerged vegetation in a stream; the stonefly larva *Taeniopteryx* and the larva of the haliplid beetle have backwardly directed spines along their bodies.

Glancing casually at the stream bed we
may come away with the impression
that few animals live there. It is not
until we take a net and prod about
among the stones and finer sediments
that a wealth of animal life is revealed.

Hiders under stones

There are very few stream-dwelling
invertebrates that do not spend at least
part of their life cycle hiding beneath
stones. A number of factors may be
involved, either singly or in
combination. Beneath the stones the
animals find protection from the
current; they may avoid predators; they
may be negatively phototactic (light-
avoiding); they may be positively
rheotactic (crevice-seeking); or they
may feed on decomposing organic
matter which becomes trapped beneath
the stones.

There is no doubt that the
environment beneath the stones is far
less physically demanding than life on
the tops or sides of the stones. Beneath
stones the speed of current flow is very
slow indeed or even non-existent; in the
latter case the 'dead' water tends to be
very poorly oxygenated and is not
widely exploited. The majority of
bottom-dwellers do require some slight
current, where the light, temperature
and oxygen environments tend to
remain fairly uniform. It is under such
stones that many of the animals which
show few adaptations to life in a current
are to be found. Typical of such animals
is the freshwater shrimp *Gammarus*.
This animal, although flattened from
side to side, is not particularly
streamlined. Some of its limbs end in
claws which enable it to cling weakly to
the bottom, and it is capable of
swimming in an upright position
through the water, though not for long,
nor against the current. Well over 90
per cent of the shrimp population may
be found beneath stones during
daylight hours, though this is not the
case at night (pp. 80–1).

Besides the non-specialist animals,
many of those highly adapted to life in

fast-running water will retreat to the
relative safety of the under-surface of
stones in times of flood or when some
physiological processes make it
desirable. For example, the larvae of
some cased caddis flies and the blackfly
Simulium are found most commonly on
the tops and sides of stones, but when
they are about to pupate they migrate to
the bottoms of the stones where they
firmly attach themselves and complete
their metamorphosis. Similarly, many
organisms retreat beneath stones for
short periods when they are about to
undergo a moult, as immediately after
moulting their bodies are extremely soft
and vulnerable to attack. Some
predatory animals also hide beneath
stones. One fish, the bullhead, hides in
its lair and rushes out to seize the
animals carried towards it by the
current.

In addition to the pockets of 'dead'
water found beneath stones, similar
areas are created among clumps of
water weeds, and these often house very
large numbers of invertebrates. The
spine-like processes on the backs and
sides of the bodies of haliplid beetle
larvae, the nymphs of the mayfly
Ephemerella and the stonefly
Taeniopteryx may help them to stay put
among the weeds.

Burrowers in the mud

One of the animals most highly adapted
to a burrowing mode of life is the cream
and brown coloured nymph of the
mayfly *Ephemera*. It constructs its
burrows in areas where the substratum
is composed of fine sand and silt, using
the modified tusk-shaped mandibles, or
jaws, which project some way in front
of its head. These are pushed into the

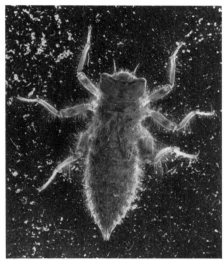

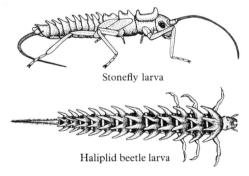

Stonefly larva

Haliplid beetle larva

soft sediments. The flattened, hairy forelimbs are forced forwards, one on each side of the mandibles, and then pushed out sideways, widening the gap between them. The nymph then rams its head into the gap thus created and the process is repeated until a burrow some 4–6 cm long is constructed. This may go deep down into the sediments, or it may lie almost parallel to the stream bed. The hairs which cover the surface of the animal's body and limbs buffer the weight of the sediments, preventing it from being crushed. In times of flood the animal buries itself deeper into the sediments, but under low flow conditions it may occasionally be seen poking its head out of the burrow, then quickly withdrawing back into it. This is a method of drawing oxygenated water into the burrow. The finely branched gills lie curved tightly

over the back and are wafted gently up and down. The action of the gills draws a current of water over the animal's body. If some Indian ink is added to the water just in front of the head, the pattern of water flow can be observed. Water is drawn from the sides of the body across the gills and then passed down the length of the body surface. This is the reverse of the flow pattern to be found in most free-living mayflies, where water is drawn along the body axis, passed across the gills and discharged at the sides of the body. In a burrowing animal this lateral discharge of water would lead to the erosion of the sides of the burrow. *Caenis* is another mayfly found in silty areas of the stream. It does not construct a burrow but lies partly submerged in the soft sediment. Its gills are protected by a horny flap (in fact the second gill) beneath which the remaining gills may be seen flickering rapidly. Another animal which partly buries itself in the stream bed is the hairy nymph of the dragonfly *Cordulegaster*. The hairs on the upper parts of its body trap silt and organic detritus and it rapidly takes on the appearance of the soft sediments round about it. There it lies in wait for unsuspecting prey to move within striking distance. Then the 'mask'—the modified fused third pair of jaws ending in strong hooks—is shot out and the prey is seized by the hooks before it is eaten.

The art of camouflaging the body to make it blend in with the surroundings is used by a number of animals to hide from their potential predators. The larvae of some caddis flies are such perfect masters of disguise that it may be some time before their cases can be distinguished from a piece of twig or a dead leaf lying in the bottom of a sorting tray. Other animals hide elsewhere. Some midge larvae, or chironomids, spin a silken tube about them which traps silt and detritus on its surface. The chironomids are a particularly successful freshwater group: well over four hundred species are to be found in the British Isles alone. They have developed a number of ways of avoiding the current and predation. Some (the so-called blood-worms) contain the red oxygen-carrying pigment haemoglobin and are able to lie partly buried in the soft sediments of the most organically rich streams and lakes. Others bury themselves in the leaves, stems and roots of aquatic plants. The larvae of some *Chrysomelid* beetles and the maggots of the fly *Chrysogaster* also find refuge among plant roots. The small horny tube at the end of their bodies is pushed into the air-transporting tissues of the plant so as to obtain a ready supply of oxygen while lying perfectly protected from the rigours of the outside world.

PROJECTS

1 Pick up a flat, hand-sized stone from a stream. Very quickly scrub the organisms off the bottom surface into one dish and place the stone with the remaining animals into a second dish. Repeat, sampling 20 stones. Identify and count the organisms removed from the bottoms of the stones. Record your results. Scrub off, identify and count the organisms remaining on other surfaces of the 20 stones. Estimate the percentage of each species living under the stones using the following formula—

$$\frac{\text{No. removed from bottom of stones}}{\text{No. removed from bottom + no. removed from top and sides}} \times 100$$

Are freshwater invertebrates found more commonly on the tops and sides or on the bottoms of stones? Is the distribution pattern the same at night? (Use a torch covered with red cellophane.)

A stone scrubbing experiment in a hill stream.

2 Set up a tray as shown below. Add 20 freshwater shrimps. Once every half-hour observe the number of shrimps in the illuminated half of the tray, and record. Repeat throughout the day. Do shrimps exhibit light avoidance (negative phototaxy) or are they light-seeking (positive phototaxy)? Try this experiment with other common species. Return all animals to the stream afterwards.

Upper tray covering half of lower tray

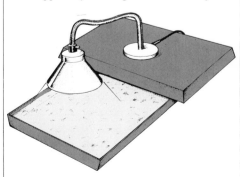

Apparatus to demonstrate whether freshwater shrimps are light-seeking or light-avoiding.

BREEDING AND BROODING

The majority of animals in fresh water abandon their eggs as soon as they are laid, but some species, including several fish, spend much time and effort looking after their eggs and even their offspring.

Spawning

Various ways in which freshwater invertebrates lay their eggs are outlined on pp. 28–9. Fish also lay their eggs in different ways. Even within a single species, the number can vary, depending on the size of the female. Most, however, spawn in shallow water; salmon, trout, minnows and lampreys choose a stony bottom in rapid currents, while bream, carp and perch prefer weedy, stagnant water. Generally, fertilization of the eggs takes place externally in the water. Bullheads and gobies, however, pair up so that the eggs can be internally fertilized.

Jawless lampreys use their suctorial mouths when spawning: the female clings to a stone and the males clamp on behind her head so that they can entwine their bodies around hers and fertilize the eggs as soon as they are laid.

The stickleback story

The widely dispersed three-spined stickleback is a good example of a fish which both undergoes an elaborate courtship display and cares for its young. At the beginning of the breeding season, in spring, the males isolate themselves from the school to select their territories. They also develop a spectacular, coloured breeding dress, with blue eyes and a bright red throat and belly.

Each male then builds a nest, first excavating a shallow pit by removing mouthfuls of sand from the bottom. Bits of algae and weeds are gathered and cemented together with a sticky kidney secretion, and the fish then makes a tunnel by wriggling through the mass of plants. When the nest is finished, the red belly becomes even brighter and the back becomes a shiny, whitish blue. If a male in breeding dress encounters another, he adopts a threatening posture, darting towards him with an open mouth and erect dorsal fins.

Meanwhile the bellies of the female

▲ A pair of three-spined sticklebacks *Gasterosteus aculeatus* prior to breeding, when the male (below) is sporting his red belly and blue eyes.

◀ Courtship of the three-spined stickleback *Gasterosteus aculeatus*.

▶ The female bitterling *Rhodeus amarus*, with her elongated ovipositor, approaches a swan mussel chosen by her mate.

▶ A freshwater crayfish *Austropotamalius pallipes* rears up against the aquarium glass, revealing the eggs attached beneath the abdomen.

▶ Brook lampreys *Lampetra planeri* spawn in gravel-bedded streams in spring. At the top left of the group, a female has attached her sucking mouth to a stone, and two males have entwined their bodies round hers.

sticklebacks have become swollen with developing eggs. When a male is ready to receive a female, he performs his zig-zag courtship dance, enticing her to follow him down towards the nest. After a while, she enters the nest, where she stays with her head and tail emerging from the ends. Once she has spawned inside the nest, the male enters to fertilize the eggs.

The male may court more than one female and thereby have several clutches laid in his nest. It is he who then cares for the eggs by chasing off intruders and by ventilating the eggs as he fans a water current down on to the

nest. After seven to eight days, the eggs hatch and the male continues to look after the young fish for some days by gathering them up into his mouth if they move too far away from the school.

The breeding dress of the male ten-spined stickleback is black. This colour difference, as well as variations in the behaviour pattern, ensures that interbreeding between the two species very rarely occurs.

Bitterlings and mussels

Perhaps even more remarkable is the life of the bitterling, during which the presence of a freshwater mussel is

▼ The lacy egg ropes of perch *Perca fluviatilis* are laid in weedy, slow-flowing water, where they adhere to plants under water. 200,000 eggs are produced per kg of body weight.

essential. In the spring, the female develops a 6cm-long fleshy ovipositor. On finding a swan mussel, the male fish which develops a breeding dress of reddish fins, yellow belly and white tubercles on his head, stands guard over it so that the female can insert her ovipositor inside the shells to lay her eggs. These are fertilized when the male ejects his sperm into the inhalant current set up by the mussel's gills. The mussel appears to suffer no harm from the presence of the fish eggs; indeed, it may well gain if the fish acts as a host for its own parasitic larvae (pp. 28–9). The fish clearly benefits by the protection of the eggs and young larvae.

Care of young

Besides sticklebacks, other fish which care for their young include the bullheads and the freshwater blenny. These are related to the marine sea scorpion and blenny (or shanny) respectively, which also show parental care (pp. 144–5). The female bullhead, or miller's thumb, lays about a hundred sticky eggs beneath a large stone, where they are guarded by the male until they hatch about a month later. The freshwater blenny occurs in fresh water along the Mediterranean coastline and, once again, the male guards the eggs.

The practice of carrying both eggs and young is widespread among leeches. During the sampling of freshwater life with a net in summer, it is not uncommon to find a leech with a cluster of several young on its underside. As the parent leech elongates to move forward, the young appear as a dark swelling. Since leeches are hermaphrodites (pp. 28–9), all individuals of species which brood their young will carry offspring.

In freshwater crayfish, however, it is only the female which carries the young. When she mates, in September to November, the male leaves his sperm capsules near the entrance to the oviducts, so that as the 50–350 eggs are laid a few weeks later they are immediately fertilized. These eggs are then carried around by the female beneath her abdomen for a further six months before the larvae hatch and are carried for two more weeks.

Why brood?

In the battle for survival there are two main types of reproductive strategy which species have adopted in the course of evolution. Most species that live in habitats which are short-lived on a geological time scale have a weed-like solution to ensure that the species continue to exist. They have an annual life cycle during which a large proportion of their energy is expended on producing myriads of tiny eggs in a single reproductive effort. Each minute egg contains a minimum investment of energy, because in the grim gamble for life the chance of survival is almost non-existent. It is rather like doing the football pools; the more bets you have, the better the chance of winning – the prize being survival.

The other strategy is to live longer, and to reproduce repeatedly, producing only a few eggs or live offspring, each with a big investment of energy. When the young are brooded internally, or carried around, or cared for after hatching, the dice are further loaded to help ensure the survival of the young. The most advanced expression of this trend is the life cycle of man himself.

PROJECTS

1 Set up a freshwater aquarium in the spring. Anchor some weeds with stones and gravel collected from a stream. Collect a male three-spined stickleback with a red belly and put him in the aquarium. Watch to see how he builds his nest.

2 Attach a small rectangular mirror to the outside of the glass aquarium with adhesive tape so that the stickleback can see his reflection in it. How does he react to seeing what he assumes to be another male in his territory?

Remove the mirror. Cut out an oval piece of brown card approximately the same size as the stickleback. Paint one part with non-toxic red paint. How does the stickleback react when this model is stuck to the outside of the aquarium?

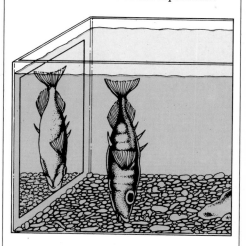

A male three-spined stickleback *Gasterosteus aculeatus* adopting a threatening posture in response to his own reflection in a mirror.

A male three-spined stickleback threatening a model of another male stuck to the outside of the aquarium.

3 Remove the model and introduce a female stickleback to the aquarium. How does the male react to her presence?

ANIMAL ARCHITECTS

Many animals living in fresh water secrete substances or build structures around themselves in order to increase protection or gain a home, a site of attachment or a method of capturing food.

Safe residences

If a fine net is used for sampling a peaty pond, the net contents, when examined under the microscope, may include some single-celled organisms, or protozoans. Some of the simple, one-celled amoebae, such as *Arcella*, secrete leathery, cone-shaped houses around themselves. They look rather like miniature limpets, and from beneath the shell finger-like pseudopods emerge, moving the animal along and engulfing surrounding bacteria and algal cells on which they feed. *Difflugia*, another 'cased' *Amoeba*, is protected by a leathery coat covered with sand grains, often with four small castellations at the corners. In ponds and lakes, scrapings from the under-surface or stalks of lily pads will often contain the microscopic cylindrical homes of the rotifer *Floscularia*, whose sand grain cylinders are firmly cemented at the base to the plant. At the free ends, a crown of microscopic, hair-like cilia may be seen in constant motion, sweeping suspended particles into the mouth.

Interior designers

Food-bearing currents can also be observed near freshwater sponges if a small quantity of Indian ink is added to the water above them. These rather uninteresting cream or green animals, which may be found encrusting the sides of disused canals or of jetties in some lakes, are in fact gems of internal architecture. They have a large number of cell types, each with its own specialized function, supported on a scaffolding of intricately shaped and angled siliceous rods, called 'spicules'. If the sponge is forced through a fine sieve, the cells will reform into a complete sponge after several days.

Caddis constructions

Of all the freshwater animal architects, the cased caddis must be among the most familiar. The British freshwater biologist Dr Norman Hickin, who has made a lifelong study of caddis, has suggested that these insects may be

◄A variety of caddis cases made of twigs, stones, or neatly shaped pieces of vegetation.

►A close up of a caddis larva *Phryganea* sp feeding on ivy-leaved duckweed *Lemna trisulca*. The case is made of pieces of leaf, cut and rolled around each other.

▼Caddis larvae *Limnephilus vittatus*, emerged from their sand grain tubes, feed on water moss *Fontinalis*.

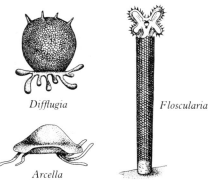

Difflugia

Floscularia

Arcella

▲Protozoan architects that carry their homes on their 'backs', *Difflugia* with a helmet made of sand grains, and the leathery, limpet-like cap secreted by *Arcella*. The tube-like sand grain cylinder of the rotifer *Floscularia* is firmly attached to a plant at the base.

◄A flattened case made of beech leaves conceals a caddis larva *Glyptotaelius pellucidus*. The emerging caddis reveals the gills that keep a current of water passing through the case.

▼The water spider *Argyroneta aquatica*, with its partly completed silken air bell attached to submerged plants. Both the adults and their offspring tap this underwater air supply.

named after 'caddis-men' – the travelling vendors of ribbons, cloths and braids (caddis), which they pinned to their coats.

Caddis larvae use one of two main building methods to construct their cases. The majority begin by spinning a silken cylinder around their body, the threads being produced from the salivary glands in the mouth of the larva. To this cylinder, the building blocks – stones, sand grains, empty snail shells, twigs, leaves or stones – are cemented. A pair of hooks attached to the last abdominal segment latches into the walls of the cylinder, which is then dragged along as the larva moves around. A few caddis use a different method of case construction: the larva walks over a sandy section of the stream bed spinning silk threads over it. When a carpet of threads has been produced with the sand grains incorporated, the larva abandons its old case and quickly rolls itself up in the new one, the overlapping seam of which is 'sewn' up!

As well as protecting the soft bodies of the larvae, the cases

camouflage them. Heavy cases help caddis larvae to maintain their position in fast-flowing streams, whereas cases made from the leaves of Canadian pondweed aid buoyancy for some pond-dwelling caddis. On warm summer days, they can be seen grazing the algae off the surfaces of other pondweed plants, trapping the bubbles of oxygen given off by them in the entrance of their cases. Once sufficient oxygen has been collected, a caddis floats up to the next strand of pondweed and starts feeding again. If it floats to the surface, the larva simply pushes out the bubbles and sinks down to the pond floor.

Caddis larvae are usually very conservative in their choice of materials for case construction. However, some species of *Limnephilus* may start their cases using plant remains, then suddenly shift to other materials, such as small stones or snail shells, and then back to plant remains at a later date. The gills of cased caddis are generally found on the backs of the animals. In poorly oxygenated waters the larvae may be seen undulating their bodies, thereby drawing a current of water into the case, over the gills, and out through the rear of the case.

Some caddis do not build cases but live within a net which they spin, attaching its walls either to the sides of stones or to the mosses growing on them (pp. 78–9). The occupants tend to sit in the net with their heads facing downstream, as in this position they are able to graze the algal cells and small invertebrates that are trapped by the net. They move backwards out of the net using hooks situated on the long false legs at the hind end of the abdomen. When the larvae are ready to pupate they may cement a few stones to the net, thereby creating a very crude case. Some chironomid midge larvae create tube-like cases to which silt may become attached, while some spin silken cocoons in which to live. These may be attached to stones or plant surfaces, or may lie just beneath the bottom surface in soft muddy areas.

Homes and nurseries

The larvae of china mark moths also build a protective case. They live beneath waterlily and pondweed leaves, inside a case made from pieces of leaves attached together by silk. *Cataclysta lemna* is a smaller caterpillar which makes its case from several duckweed plants.

In ponds, two other forms of home can be found: the silvery diving bell of the water spider (pp. 22–3) and the leafy nest of the stickleback (pp. 74–5). The lodges built by the muskrat and beaver (pp. 88–9) are much grander constructions, which involve a great deal of effort in the transportation of materials to the lodge site.

(pp. 78–9). (pp. 22–3) (pp. 74–5). (pp. 88–9)

PROJECTS

1 Collect 20 cased caddis larvae from a pond. Examine their cases, drawing different types and recording what proportion are constructed from live plants, dead leaves, twigs, sand grains or shells. Return the caddis to the pond and repeat with 20 caddis larvae collected from a stream. Make sure to return them to the stream. Compare the results. Which type of case blends in best with its natural surroundings? Is this the most abundant type?

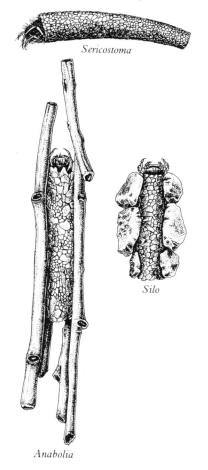

Sericostoma

Silo

Anabolia

2 The cased caddis *Anabolia* is easily recognized by the very long twigs attached to the back of its case. Collect a number and drop them into a slow-flowing stretch of stream. How do the cases align themselves to the current? How is this an advantage to the animals when they feed?

3 Collect some duckweed and search for cases of the moth *Cataclysta lemna*. How does the case construction differ from caddis cases made from living plant material? Examine the undersides of waterlily pads for larger cases of *Nymphula nympheata* larvae. What proportion of the pads are used by this larva as a shelter?

PREY AND PREDATORS

All life is dependent on plants, either directly as herbivores or indirectly as carnivores. In all food pyramids, it is the plants which are most numerous and form the base of the pyramid.

About 40 per cent of the food that the plants (primary producers) create is subsequently consumed by the herbivores (primary consumers), some of which are themselves eaten by carnivores (secondary consumers) which in turn may be preyed upon by other animals higher up the food chain (tertiary consumers). Finally, the top carnivores prey on animals lower down the food chain, but are themselves rarely, if ever, preyed upon.

The primary producers

The interrelationships between predators and prey are readily studied in a small upland hill stream. Here, the large flowering plants (macrophytes), so characteristic of slow-flowing lowland rivers, may be absent, uprooted by sudden flash floods occurring in winter. The larger green thread-like algae may also be scarce, so the most important photosynthesizers (pp. 42–3) are the myriad microscopic algae, the diatoms and the desmids which coat the stones and coarse sand grains of the stream bed. The abundant diatoms when seen under the microscope appear golden brown instead of green, because the green pigment chlorophyll present in all algae is masked by the brown pigment phaeophycin. Diatom cell walls are impregnated with silica which is absorbed from the surrounding water, producing complex and beautiful shapes (pp. 40–1). Photosynthesis is also carried out by mosses and liverworts growing on the surface of the stream boulders, and by the bankside plants which dangle into the stream.

The consumers

The ways in which herbivores feed on the plant tissue are many and varied. The river limpet, for example, like its marine counterpart, uses a horny, file-like tongue, or radula, to rasp algae from the stones. Many of the stream-dwelling herbivores are insects. Their mouthparts usually consist of a pair of jaws, or mandibles, which have rather broad, rounded cutting edges, that are ideal for crushing the siliceous cell walls

▲One of the detritus feeders, the water hoglouse *Asellus aquaticus*, which helps to recycle the nutrients of the stream or pond.

▶The brown trout *Salmo trutta* is a tertiary consumer in the stream food webs, for this mature fish will feed on a variety of organisms, including smaller fish.

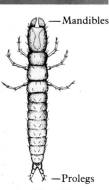

— Mandibles

▶A net-spinning caddis *Philopotamus montanus* taken out of its net to show the basal prolegs that maintain its position in the net. Although it has large mandibles, it is a herbivore, and the rounded edges of these are used for crushing plant cells.

—Prolegs

Mandibles

▶The alderfly *Sialis lutaria* is a free-swimming predator with twelve feathery gills along its abdomen. The prey is caught in the jagged mandibles, which also cut it up.

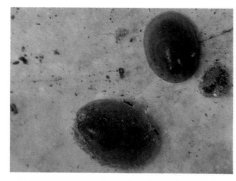

▼River limpets *Anacyclus fluviatilis* are able to hang on to a stone and graze the algae on its surface with their file-like tongues.

▲This colonial protozoan *Carchesium* also aids recycling of nutrients within the freshwater system by further breaking down the organic matter.

of the diatoms. As well as the mandibles there are usually two finger-like processes – the palps – which help to direct the food into the mouth.

In some stonefly nymphs, the palps are armed with bristles which help to sweep the algae off the stones, while in some mayflies stout spines serve the same purpose. A number of caddis flies supplement the action of the palps by using the forelimbs. The algae which these displace are swept backwards on to long hairs which fringe the second pair of limbs. These are then grazed clean once sufficient algae have accumulated on them.

The blackfly maggot *Simulium* also uses hairs for feeding. These are attached to the palps and form two great fans, or feeding combs, one on each side of the head, which are used to filter out algae and bacteria carried in suspension by the current. The adult females of some blackflies feed on birds and mammals; *Simulium tuberosum*, known as the birch fly, plagues human visitors to the Scottish Highlands. Filtration is also used by other organisms to catch their food; notably the net-spinning caddis flies spin funnel- or cigar-shaped nets which face into the current and catch particles in suspension. In small upland streams two groups of caddis fly are commonly encountered: *Philopotamus* near the head waters, and *Hydropsyche* in the middle reaches of the stream. *Philopotamus* is a herbivore which builds a net fine enough to catch

the single-celled diatoms and strong enough to withstand the turbulent flow of the headstream waters. The nets of *Hydropsyche* are of a larger mesh but made of finer threads, and, although it is a more carnivorous feeder than *Philopotamus*, it competes with it for sites in which to spin its net. It was previously thought that their ecological separation in the stream had come about through competitive exclusion, but there is now evidence to show that their distribution is controlled by temperature rather than competition.

Another sedentary carnivore is the dragonfly nymph *Cordulegaster*, which lies partly buried in the sediments of the stream bed, with its head, large eyes and short antennae protruding. As any suitable prey drifts past, the nymph shoots out its mask and grabs the prey with a pair of toothed, pincer-like processes. The mask (a horny structure made of the backward-pointing lower lips and fused palps) is then withdrawn beneath the head and the mandibles get to work. These are typical of most carnivorous insects, and have narrow cutting edges which work with a slicing action rather like a pair of scissors.

Most carnivores actively seek out their prey, roving from stone to stone, using their large eyes and, in the case of the large stonefly nymph, their sensitive antennae, to detect prey. The cranefly larvae *Pedicia* and *Dicronota* seem to have fewer organs to help them locate their prey, their head capsule consisting of a large pair of jaws, a small pair of antennae and very little else.

The larger carnivores, particularly the fish, are capable of rapid bursts of speed in pursuit of their prey. The bullhead shelters beneath stones from which it makes rapid forays into open water. The trout, on the other hand, can be seen poised almost motionless in the open water, keeping a fixed station over the stream bed, using a rock on the bottom or a bush on the river bank as a marker, and waiting for its prey to drift towards it.

Uneaten plants and animals grow, reproduce and eventually die. Their remains and any waste products that they have produced during their lifetime are rich in nutrients and these are utilized by a group of animals, the detritus feeders, which includes worms, bloodworms, freshwater shrimps and water hoglice, as well as by microscopic decomposers such as bacteria and aquatic fungi.

In addition to providing food for aquatic carnivores, the stream is a valuable source of food for terrestrial visitors, such as the dipper, kingfisher, heron and otter.

Although the combination of species inhabiting lowland streams is different, similar interrelationships between prey and predators exist.

PROJECTS

① The freshwater shrimp *Gammarus* is most commonly thought of as a detritus feeder. Collect a large number of shrimps and divide them into two large jam jars filled with stream water which has been filtered through a clean handkerchief. Add fresh green leaves of aquatic plants, mosses or algal filaments to one jar, and decomposing leaves from the bottom of a pond (first washing them to remove other organisms) to the other. Which group survives for the longer period? Observe the shrimps carefully and see which of their many appendages are involved in feeding.

② Lift a number of large stones from the bed of a clean hill stream to find the nets of caddis flies *Philopotamus* and *Hydropsyche*, which appear as a rather silty series of fibres to which smaller stones and sand grains may be attached. Take these home and gently brush the contents out of the nets with a paint brush on to a glass slide. Observe the contents under the microscope and compare the catches made by these two caddis.

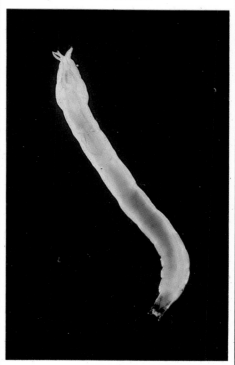

Pedicia rivosa – a tipulid larva from a hill stream

③ Collect a number of cranefly larvae (*Pedicia* or *Dicronota*) and feed them on dyed fly maggots (obtainable from fishing tackle shops). How long does it take for the food to pass along the gut of the animal? Does the time increase or decrease if you place the jar of water in a sunnier position? What is the function of the silver tubes running the length of the larvae's bodies?

THE STREAM AT NIGHT

A casual daytime glance at the stream bed reveals relatively little animal activity. At night, things are very different.

Simple investigations show that the majority of freshwater invertebrates are to be found beneath stones during daylight hours. There are some animals, however, such as the river limpet, the blackfly larva *Simulium* and some caddis larvae, which are exceptions to this general rule (pp. 70–1). There are also seasonal variations; the young nymphs of many insects are so small when they first hatch that most of their body surface lies within the boundary layer and they may be found in great abundance on the tops and sides of the stones. As they moult and grow larger, so they begin to seek the under-surface of the stones.

Herbivores form the most numerous group in the majority of animal communities. In many streams and rivers there are relatively few higher plants (macrophytes), and microscopic brown and green algae (diatoms and desmids) are the primary source of food. These are largely distributed on the tops and sides of the stones where sunlight can reach them. This creates an interesting spatial separation between the herbivores and their food supply and they must move from beneath the stones to the tops and sides at some time during the day or night in order to feed.

Invertebrate drift

Under non-turbulent conditions water exhibits laminar flows; that is, it behaves as if it were made up of flat sheets, one above the other. Those nearest the surface flow most rapidly; those closest to the bottom are slowed down by friction. As the animals move up towards the water surface to feed, there is a greater likelihood of their being swept away by the current, and becoming part of the phenomenon known as 'invertebrate drift'. There are two main types of drift – voluntary and involuntary. In the former, the animal voluntarily lets go of the substratum and is carried off downstream. This might be the response of an animal living near the head of a stream which is drying out in the summer months, or of an animal occupying an area where population densities are becoming too

▲A drift net in position in a stream.

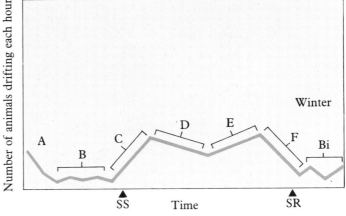

Typical summer and winter patterns of invertebrate drift. The lettered phases are referred to in the text.

SS = Sunset

SR = Sunrise

▶A stony hill stream, showing the riffles and pools that develop along its length.

high, or where the pattern of current flow changes, bringing less food to it.

There are several types of involuntary drift. The most important are casual drift, which is caused by the stream flow eroding the stream bed and washing a few animals away with it; catastrophic drift, which is the large-scale movement of the bed and invertebrates when the stream or river is in flood; and behavioural drift, a

consequence of the fact that when an animal moves it reduces the number of limbs in contact with the substratum, which increases the risk of its being swept away by the current.

The animals drifting downstream may be collected in a simple drift net. If this is sampled hourly throughout the day and night it is possible to learn something of the behavioural patterns of animals living in the stream.

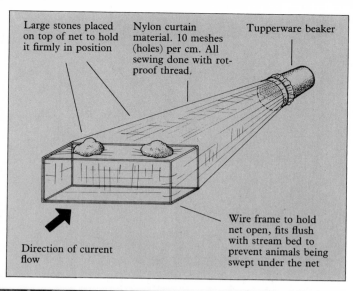

▶An invertebrate drift net used to sample movement of invertebrate fauna in a stream at different times of day and night. A large nylon net (10 meshes per cm) is sewn round a metal or strong wooden frame with rot-proof thread, and a plastic beaker is attached to the end with waterproof tape. This is placed in the stream, so that the frame fits flush with the stream bed to prevent animals from being swept under the net, and the top is weighted with heavy stones to withstand the current.

Large stones placed on top of net to hold it firmly in position

Nylon curtain material. 10 meshes (holes) per cm. All sewing done with rot-proof thread.

Tupperware beaker

Direction of current flow

Wire frame to hold net open, fits flush with stream bed to prevent animals being swept under the net

Patterns of drift typical of warm sunny nights and of cooler winter nights are shown in the graph. A number of distinct phases can be picked out in these patterns. Firstly 'A', initial disturbance, brought about by placing the drift net in position in the stream. This disturbs the bed and alters the pattern of flow. Throughout the day, there are minor erratic fluctuations in the numbers of drifting animals as a result of casual drift 'B'. At sunset the number of animals drifting shows a marked increase 'C'. Some large stones, painted a pale colour and left for at least 24 hours on the stream bed, may be observed at night, every hour or so, by using a torch with a red filter on the front. It will be seen that the increase in drift coincides with a frenzied period of activity when the animals move from beneath the stones on to the tops and

sides, that is, there is a strong behavioural component to the drift. In summer, this is followed by 'D', a steady decline in the number drifting, as an increasing number of animals, having fed, once again reach the comparative safety of the under-surface of the stones. Winter nights are longer and colder and 'C' is followed by a period when the animals slowly feed, all limbs in contact with the substratum 'D'; the drift rate falls slightly but rises again before dawn as the animals move back beneath the stones 'E', which they achieve during phase 'F'. Following the night's activities, casual drift 'Bi' becomes increasingly important once more.

Controlling factors

It can be seen from the graph that light intensity plays an important part in determining when the animals become active, that is, it controls the timing of the animals' behaviour. Differences in the shape of the summer and winter curves suggest that temperature is important in determining how many animals take part in the drift, that is, it controls the amplitude of the drift rather than the periodicity. Other factors affecting the amplitude of the drift include the amount of water coming down the stream (discharge) and the density of animals on the stream bed.

In addition to these external triggers there are internal biological clocks ticking away inside each animal's body which help to determine the periodicity in their behaviour. These are largely under the control of a series of hormones which are released from the brain. Early in the morning there are few neurohormones stored in the brain of the animal. However, the hormones build up steadily in the mid brain throughout the day until about an hour before the animal is due to become active (as light intensity is falling). They are then released from the brain into the blood and carried throughout the body, bringing about a state of 'readiness' within the animal. Once it has become sufficiently dark the animal becomes active. The time at which this occurs changes throughout the year as day length increases and decreases with the seasons, and because of this the animal does not become active at exactly the same time on each successive night.

Patterns of activity which change with increasing and decreasing day length are described as 'circadian' rhythms (Latin circa, about; dies, a day).

The invertebrate drift is exploited to a greater extent by visual hunters than by those which are attracted by smell. Thus most visual hunters, such as trout, will be largely crepuscular, feeding at dusk and dawn.

THE FERTILE FRINGE

There are still remnants of ancient habitats to be found at the edge of a stream, supporting a variety of animals and plants that are masters in the art of transformation.

Each small stream winding its way through the landscape is like an artery that carries food, minerals and oxygen in its waters. This supply constantly travels on through moorland and meadow past clumps of watercress and brooklime, between lines of alders and willows, gathering more water from other streams, until it is a vast, silt-laden river making its way to the sea. Yet this continuous ribbon of water is isolated from other streams in the next valley, so that any organism that lives here must have a means of dispersal to a neighbouring stream in another catchment area. This means using a medium other than water at some stage in the life cycle.

Between land and water

Each winter the fringe of green at the edge of the stream may disappear under flood water scouring out the channel or burying the shoots under new deposits of sand or gravel. Yet any small fragment of stem that is left has the capacity to grow into a clump when conditions improve, rapidly re-invading the new deposit with shoots that root at every node. Within these clumps the current is slowed down and finer sediment dropped, creating another habitat for organisms that require a sheltered place to live and feed. The tall monocots—grasses, sedges, rushes and iris—have another method of surviving the difficult times. Their stems are deep underground in the stream silts, and in the event of damage new shoots appear from the rhizome. These underground stems form a close network, rather like wire mesh, which requires enormous water force to remove it. Their upright stems also trap silt and slow the current, and these may entirely change the character of a stream, filling in the channel so that the water is forced to find another route in a winter flood.

Between water and air

In summer the stream is full of vegetation that has emerged above the

▲A small stream that makes its way through lowland meadows supports a bank flora of alders *Alnus glutinosa*, meadowsweet *Filipendula ulmaria*, hemp agrimony *Eupatorium cannabinum*, some tall, narrow-leaved emergents of branched bur reed *Sparganium erectum*, and lush summer clumps of brooklime *Veronica beccabunga*, watercress *Rorippa nasturtium-aquaticium* and Fool's watercress *Apium nodiflorum*.

▶An eyed hawk moth caterpillar *Smerinthus ocellata* is perfectly camouflaged against the willow foliage on which it is feeding.

▶When a large batch of Grannom caddis flies *Brachycentrus subnubilus* emerges from the water in May, many get trapped on the sticky hairs of comfrey plants growing by the water.

▼The bean galls on this willow leaf are caused by a small sawfly *Pontania proxima* that lays its eggs in the tissue of the leaf, which then produces a gall that is specific to the insect.

▲Policeman's helmet *Impatiens glandulifera* is a native of the Himalayas but has spread along stream margins elsewhere by means of explosively released seeds.

water surface, each stem jostling for a position in the sunlight, and July brings tall spikes of flowers on every aerial shoot. Although the abundance of insect life above the stream provides plenty of agents for insect pollination, many of these plants must rely on the stream for seed dispersal. Policeman's helmet is a native of the Himalayas, but has rapidly colonized stream edges in other parts of the world. The showy pink flowers are pollinated by bees, but the seeds are explosively released from a pod into the margins and the current whence they may be carried to new places.

While plants may show adaptations to underwater and aerial life on the same shoot, the emerging insects must adopt a new method of transport and respiration in the air, and this takes time to accomplish. The plants provide the route by means of which the aquatic dragonfly nymph may climb out of the water and undergo the final transformation to adult life leaving its last nymphal skin on the stem. From a soft-bodied, crawling bottom-dweller it becomes a horny-bodied, transparent-winged creature whose only connection with its former life is the plants on which it alights. It can now travel to new streams, and hunt over damp meadows—but has always the final problem of returning the next generation to the water. Each species of dragonfly and damselfly has adopted its

own method: some fly low over the water, dipping their abdomens in the water to release the fertilized eggs; others crawl down the stems of the fringe plants into the water, encased in a silvery coat of air trapped in the fine hairs on their bodies, and lay their eggs below the surface on the stems of the plant. The alderfly has adopted another method. The larva crawls to the edge of the stream and pupates in the mud, whence it emerges as a dull, brown-winged adult. After mating the female alderfly lays her eggs on the leaves of plants overhanging the water, so that when the young hatch out they drop straight in.

Food for others

Emergence is the most vulnerable time in the insect's life, and there are many predators, particularly birds, to take advantage of it. Some species emerge at night so that their wings have hardened by dawn, but many fall prey to the birds that visit the stream to find food for their hungry youngsters. Other birds, like the coot and moorhen, inhabit the stream throughout the year, feeding mainly at the surface and among the fringing vegetation. The water vole is also a resident of the stream banks, feeding on the shoots and underground stems of the reeds and other water plants. Although it needs a burrow or reed nest to breed and live in, its shape and long, waterproof guard hairs make it an adept swimmer.

Waterside trees

Along the edges of our streams the remnants of the once widespread, marshy woodland are found. Agricultural development by way of cutting, draining and reclaiming of these areas has confined marshland trees to the edges of streams, where the willows and alders flourish along with all the animals that feed on them. Look for leaves that have been nibbled and you will find some of the grazing caterpillars. Some hide away in the day, sewing themselves into a leaf tube, while others, like the eyed hawk moth, are perfectly camouflaged with their surroundings, their disruptive markings making them inconspicuous among the leaves on which they feed.

The bean galls on sallow are caused by a small sawfly that lays its eggs in the leaf, which then produces a gall. Later the adult fly emerges from the gall. A great variety of galls are found on the willow, each species stimulating the leaf to produce a characteristic shape. Willows also come in all shapes and sizes, many local varieties being perpetuated by their facility to grow from broken shoots. Man has for over 2,000 years used these to grow withies, which are still harvested annually in some places to make baskets.

PROJECTS

1 How many different species of invertebrate fauna can you find living on one willow tree? Record their height above ground and their aspect when you find them. Do some species choose particular situations?

2 In early spring find a small stream with fringe plants (watercress, fool's watercress, brooklime or water forget-me-not) growing in it. Mark a shoot on the edge of the stream with a stick or wire that is at least 30cm above the surface and fasten a coloured label to it. Record the rate of growth of the clump during the summer. Does it develop faster upstream or downstream? Are the emergent leaves and stems different in shape and character from the submerged leaves and stems?

3 It has been shown that birds tend to eat fairly constant numbers of insects despite fluctuating numbers available. Test this in May or June on a secluded river bank where you have a good view of the insects and a nesting pair of birds nearby. Record how many times the adults feed and what they are feeding on. What are your conclusions?

NATURAL MONITORS OF POLLUTION

Pollution is usually associated with man's activities, but how does it affect the environment for the animals and plants that live in it?

Pollution is a word we all use to cover a multitude of human activities that in some way degrade the environment, from unsightly rubbish tips to the less obvious addition of chemical and organic waste to our rivers. Because we are also dependent on fresh water we have tended to define pollution by whether it is fit to drink. But for organisms living in the stream or river the question is 'is it fit to live in?'

There are many different types of pollution that change the living potential of a freshwater site. Using water for cooling changes the temperature of the water, and warm water holds less oxygen than cold, creating a problem for the freshwater fauna. It may also affect the life cycle of the organisms that are dependent on a temperature stimulation to commence breeding or resting. Reclamation of land in a stream's catchment area causes an increase of sediment that may cloud the water and alter the stream bed downstream of the reclamation, changing the habitat for the flora and fauna of the stream bed. Chemical waste may be added by factories, changing the pH of the water as well as its mineral composition. But by far the major sources of pollution in our rivers are detergent and organic waste from domestic and farm sewage.

In nature nothing is wasted, and the primary effect of organic pollution is nutritional, causing an increased population of detritus feeders, scavengers and bacteria that break down organic material. These use much more oxygen for respiration, the oxygen level is lowered, and the stream can no longer support the populations of organisms with a high oxygen requirement such as the stonefly and mayfly naiads. The balance of the system has gone and the more sensitive animals disappear.

It might be expected that plants would be all the more luxuriant for the increased organic content in the water, but it also causes an imbalance here,

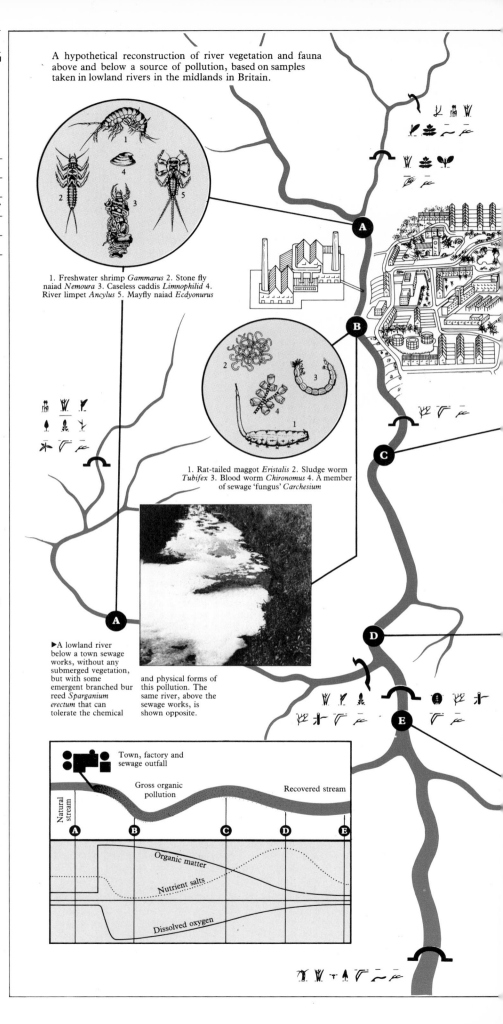

A hypothetical reconstruction of river vegetation and fauna above and below a source of pollution, based on samples taken in lowland rivers in the midlands in Britain.

1. Freshwater shrimp *Gammarus* 2. Stone fly naiad *Nemoura* 3. Caseless caddis *Limnophilid* 4. River limpet *Ancylus* 5. Mayfly naiad *Ecdyonurus*

1. Rat-tailed maggot *Eristalis* 2. Sludge worm *Tubifex* 3. Blood worm *Chironomus* 4. A member of sewage 'fungus' *Carchesium*

▶A lowland river below a town sewage works, without any submerged vegetation, but with some emergent branched bur reed *Sparganium erectum* that can tolerate the chemical and physical forms of this pollution. The same river, above the sewage works, is shown opposite.

Town, factory and sewage outfall

Natural stream

Gross organic pollution

Recovered stream

Organic matter

Nutrient salts

Dissolved oxygen

Amphibious yellow cress
Rorippa amphibia

Yellow waterlily
Nuphar lutea

Unbranched bur reed
Sparganium emersum

Reed canary grass
Phalaris arundinacea

Duckweed
Lemna minor agg

Common reed
Phragmites australis

Branched bur reed
Sparganium erectum

Hairy willowherb
Epilobium hirsutum

Fennel pondweed
Potamogeton pectinatus

Reed sweet-grass
Glyceria maxima

Water plantain
Alisma plantago-aquatica

Water forget-me-not
Myosotis scorpioides

Canadian pondweed
Elodea canadensis

Arrowhead
Sagittaria sagittifolia

Grass
Agrostis stolonifera

Club-rush
Schoenoplectus lacustris

Fool's watercress
Apium nodiflorum

Brooklime
Veronica beccabunga

Lesser pond sedge
Carex acutiformis

Tubular green algae
Enteromorpha

Blanket weed
Cladophora

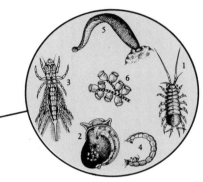

1. Water hog louse *Asellus* 2. Wandering snail *Limnaea* 3. Alder fly larva *Sialis* 4. Blood worm *Chironomus* 5. Leech *Erpobdella* 6. A member of sewage 'fungus' *Carchesium*

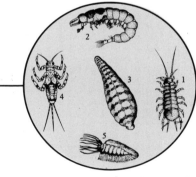

1. Water hog louse *Asellus* 2. Caseless caddis *Hydropsyche* 3. Leech *Glossiphonia* 4. Mayfly naiad *Baetis* 5. Buffalo fly pupa *Simulium*

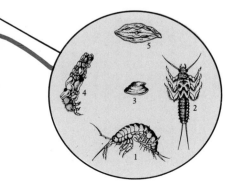

1. Freshwater shrimp *Gammarus* 2. Stone fly naiad *Nemoura* 3. River limpet *Ancylus* 4. Caddis larva *Stenophylax* 5. Flatworm *Dugesia*

and the more sensitive plants disappear. In farm ponds where there is organic effluent a thick algal scum often develops, cutting off the light from any submerged species. This may also happen in the slower stretches of river but here the proportion of bottom algae and blanket weed goes up. An additional effect of domestic sewage is the increase in salt content of the water, so that those species that are tolerant of it, such as fennel-leaved pondweed, become widespread in polluted rivers, while water crowfoots disappear quickly. Even the emergents are affected and species like the reedmace (often known as the bulrush) become abundant at the expense of others.

The major effect of organic pollution is to create an imbalance in the environment which changes the competitive status of the species living in it, so that a few species become abundant and those that are characteristic of the former community disappear. Thus there is always a lowering of the species diversity of a habitat when pollution occurs. Even so, the pollution in a river is diluted as it passes away from the source of pollution and as other tributaries join it, so that there is a recovery distance for each situation. The plants of a stream vary with geology, topography and management so that the effect of pollution can only be observed by taking sample points along a stretch of water above and below sources of suspected pollution as well as after tributaries have joined it. Chalk streams are usually rich in species but pollution cuts down the number of species here just as it does in less rich streams.

The fauna of a stream is dependent on the plant material that is available, although much of it is carried in the water itself, so that the effects of changing food material will not be as obvious as the change in oxygen levels caused by pollution. Most of the evidence for changing fauna will be found in the benthic organisms (those living on the stream bed). Once oxygen is in short supply only organisms that have a way of obtaining it can survive. These include the rat-tailed maggots with their telescopic breathing tube and the blood worms, which contain haemoglobin—the same substance that carries oxygen in mammalian blood. Minor sources of pollution can be detected only by comparing proportions of the different organisms in a standard sample from along your stretch of water. As organisms will also vary according to the type of deposit on the stream bed it is necessary to sample each bottom type at any sample point. In this way you can quantify the changes in the populations or organisms both at the source of pollution and at their recovery downstream.

PROJECTS

1 Carry out a water plant survey. Use a 1:50,000 scale or similar map and plan a survey of your local stream, starting as near the source as you can. In order to get relative samples as the stream gets larger use bridges as your sample points, so that you can distinguish what is growing on the bottom, as well as sampling it with a grab, if necessary. Take your sample on the upstream side of the bridge and record only plants that are growing in the water. The exact identification of the species does not matter for this exercise, but it is important to record all different species. Can you isolate sources of pollution? How are your species numbers affected when other streams join yours?

The same lowland river in farmland upstream with abundant submerged vegetation of unbranched bur reed *Sparganium emersum* and some bulrushes *Schoenoplectus lacustris* in the distance.

2 Sample your stream for invertebrate populations using a large tin as a core sampler (pp. 14–15). Sort and count each 'catch' in a white enamel bowl and record the numbers and types for each sample. How do your results compare with your plant survey?

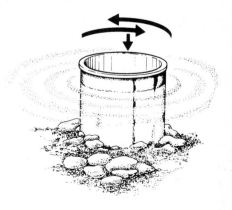

LARGE RIVERS

This broad, deep river, the Dordogne, has gathered its water from tributaries running off the uplands of the Massif Central in France. The V-shaped wooded valleys around the Massif Central, with their rushing streams, give way to cavernous limestone gorges where water gurgles from subterranean channels; and, finally, agricultural plains carry a vast river which makes its way slowly and sedately to the sea beyond Bordeaux.

The river has cut its route through the valleys over millions of years, since long before man appeared on the scene. About 85,000 years ago, early man discovered ready-made homes in the water-worn caves abandoned by the down-cutting river. Later, around 26,000 years ago, our ancestors left records of the fauna that they hunted in the valleys in the form of pictures on the walls of caves, such as those at Lascaux, deep underground. Today the wild horses, aurochs and reindeer have gone, but man has settled for centuries in numerous villages along the earliest trade route – the river. There are still about 470 km of navigable river in the Dordogne and its tributaries.

There are few fringing plants at the edge of the river below the gorges, for the winter storm flows from the highlands sweep away all but the most deeply rooted overhanging willows. But in early summer there are swathes of white-flowered water crowfoot, rooted in mid-channel on the river bed.

The Dordogne below its confluence with the Vézère, with limestone gorges in the distance.

RIVERSIDE TRACKS AND SIGNS

Some birds and mammals permanently inhabit river banks, while others visit them to drink or feed; but they all leave behind them evidence of their presence.

Since riverside mammals are mainly crepuscular (active at dusk and dawn) or nocturnal they will not easily be seen, and even those which are active by day rarely remain out in the open for any length of time. Birds which nest along river banks will move up and down their river to feed, but they never spend long in any one place. Throughout the year, however, information about the way of life of riverside inhabitants is left in the form of tracks, shed feathers and hairs, or as food remains and droppings.

Tracks

Animal and bird tracks are most likely to be left behind in soft mud, and during winter they can also be found after a snowfall. Although tracks in snow are clearly visible, they will last for only a short time before they are obliterated either by another snowfall or by a thaw. A clear track in mud rarely lasts long, before it is obscured by other tracks criss-crossing it.

Footprints are by no means the only type of track. If an animal moves along dragging its tail behind it, this will leave a continuous drag mark down the centre of the footprints; for instance, the broad tail of a beaver is quite distinct from the thin tails of the muskrat and coypu. Both adult and young otters make broad slides as they playfully descend on their bellies down river banks through mud or snow to the water. A predatory mammal may leave behind evidence of the spot where it stalked, and caught, its prey.

As well as regular river inhabitants, terrestrial animals such as foxes, rabbits, badgers and deer will move down to the river to drink in accessible places.

Tracks of birds which are most likely to be seen beside a river include those of coots (which have broad-lobed digits to their feet), moorhens, herons, swans, geese, ducks, crakes and wagtails; but, as with terrestrial mammals, other birds will sometimes visit riversides to feed and drink.

▲The weight of a swan's body makes conspicuous footprints in the muddy gravel.

▶A moorhen uses the base of coppiced trunks to make its riverside nest.

▼Soft mud shows up the track of a moorhen.

▼Walking tracks of some riverside mammals, showing relative sizes of footprints and tail track.

Muskrat

Beaver

Coypu

▲A year after a riverside poplar tree in the Camargue has been felled by a beaver, it sends up new shoots.

▼A swan mussel shell, showing damage caused by muskrat predation.

Evidence of feeding

Herbivorous animals will browse on riverside plants without scattering inedible plant remains in the way that carnivores discard mollusc or crustacean shells. Closely cropped vegetation is, however, often the best clue for finding the entrance to a water vole's burrow. Water voles will also gnaw bark in summer, as far as they are able to reach up a tree from ground level. If they debark a tree in a complete ring, they will kill the tree just as effectively as a beaver would by felling it.

The beaver, which became extinct in Britain in the fifteenth century, is Europe's largest rodent. It now occurs in Scandinavia, France (the Rhône), Switzerland (the Versoix), West Germany (the Elbe) and Poland (the Danube). Feeding mainly on the bark, shoots and leaves of deciduous trees, the beaver will gnaw right through a tree to gain access to a large area of bark. It also uses branches to dam rivers, so as to create calm water for its lodge site. Come the autumn, the beaver drags branches back to its lodge

(or burrow, in the case of the Rhône) where they are stored as winter food. Lengths of beaver sticks, with obvious gnawing marks, have been excavated from a Danish bog where the beaver last occurred in prehistoric times.

The muskrat, a native of North America, which was introduced into Europe in 1905 and is now feral, will also build a lodge, but uses reeds and rushes instead of tree branches. Muskrats tend to frequent brooks and canals with slow-moving water, where they feed on the roots and leaves of water plants as well as on freshwater crayfish and mussels.

Coypu were imported from their native South America by British and European fur farms for their underfur, which is known in the trade as nutria. Inevitably, there have been many escapes, particularly in the Norfolk Broads in East Anglia. These large rodents damage root crops by feeding, reeds used for thatching by trampling, and dykes by their tunnelling activities. In autumn and winter, water voles move into the bulb fields in the Netherlands to feed on the bulbs, while beavers enter riverside apple orchards along the Rhône to feed on fallen apples.

Both the American mink and the European mink have been bred commercially for over a century for their very valuable fur. The American mink has been repeatedly introduced to Europe and has often escaped from fur-farms. Mink are carnivores which kill more prey than they can eat, so they are a menace to nesting birds and to small mammals in the wild as well as to poultry. Their normal diet includes frogs, fish, muskrats and water voles, all of which are likely to occur near the den; but they will also forage some distance away from it. In Iceland, mink have attacked nesting ducks and waders, and in Scandinavia, salmon have fallen prey to them. A variety of corpses can often be found in their den—a hollowed-out tree or an enlarged hole of a water vole.

Bats will hunt back and forth above the length of a river for their insect food, and in so doing may drop moth wings along the river bank.

Nest sites

One of the greatest delights of a stroll along a river must surely be to see the brilliant flash of a kingfisher (p. 90) flying to or from its nest site in the bank. By laboriously pecking at the sand, a kingfisher will excavate a tunnel opening into an enlarged nest site. The burrow entrance is often whitened by disgorged fishbones. In fast-running hill streams, the curious, mossy, dome-shaped nest of the dipper may be seen built on an overhanging branch or in old brickwork.

PROJECTS

1 Be your own nature detective by looking for tracks and signs along your nearest river. Look carefully at footprints to see the number of toes, the absence or presence of claws, and the shape and size of the pads. Try to determine in which order the animals visited the area.

Beside the Dordogne river in France, a water vole leaves its tracks.

2 Make a permanent cast of a clear footprint, in the following way. Cut a piece of strong card into a strip at least 30 cm long by 5 cm deep, and bend it into a circle by sticking one end over the other. Gently push the ring into the ground around the footprint. Make up a thick, runny paste of plaster of Paris by adding the powder to a plastic container half filled with water. Carefully pour the mixture into the ring to a depth of about 3 cm. Leave it for 20 minutes to set. Gently lift up the cast by sliding a knife beneath it. Remove the outer ring and turn the cast over to see the raised impression of the footprint. Leave it overnight before washing away any mud with water. The footprint can be picked out from the white plaster cast by using a coloured paint. It can also be used repeatedly to reproduce the original footprint in wet sand or modelling clay.

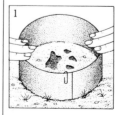
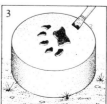
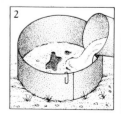

1. Pushing down mould.

2. Pouring in plaster of paris.

3. Painted cast of footprint.

Making a cast of an otter footprint.

89

RIVER BANKS

The banks are as long as the river itself, and they offer as much variety in type of habitat as does the water below them.

Many water naturalists regard river banks as an obstacle to be overcome in order to get to the river. But a closer look at the banks shows that they are an integral part of the life of the river, offering food and support for adult insects emerging from the river and breeding sites for mammals and birds that feed in or near the water but need a place on land to rear their young.

The basic material of the bank may vary from bedrock and boulders to shingle, sand or layers of mud. The stability of the bank will be affected by the action of the water below it, by the material of which it is made, and by the plants that colonize it. Where waterside trees, such as willow and alder, are allowed to grow, they may stabilize muddy and sandy banks with their network of roots. Grasses and plants which have spreading rhizomes, such as the common reed, also stabilize the banks, especially during the summer season when water levels are usually low and the erosive power of the river is much reduced. Plants with efficient rooting systems may then survive the winter storm flows. Unstable banks are often found in regions where intensive grazing and bad management both damage the bank and destroy the stabilizing plants. There are many areas however where banks are left in a semi-natural state, and where the profusion of species will delight the naturalist.

At home in the bank

When Ratty eulogizes to Mole about his riverside home in Kenneth Grahame's *The Wind in the Willows*, he describes the perfect conditions for a water vole's burrow: a muddy bank, and food close by within a 'plop' from the river in summer and winter. The burrow may get a little damp in winter, but a water vole doesn't mind that (pp. 88–9)! Most water-living animals or birds need a specific type of bank as well as a particular situation in which to make their breeding quarters, and food must be readily available from the water for the young will need feeding regularly. The staple diet of the water vole is roots and shoots of waterside grasses that colonize the banks of slow-flowing

▼The river Lyn in the south-west of England is a gentle stream in summer, where colonizers can establish themselves on the shingly winter storm flow banks to flower abundantly in this nutrient-rich and sunny spot. From left to right: hemlock water dropwort *Oenanthe crocata*, monkey flower *Mimulus guttatus*, and rosebay willowherb *Chamaenerion angustifolium*.

▲*Marchantia polymorpha* is a common liverwort of river banks and damp, shady places. The 'splash cups' contain gemmae, or vegetative units, that are dispersed by water falling on them.

▶A shaded river bank, stabilized by tree roots and colonized by shade-loving ferns, mosses and liverworts.

▼Stable and nutrient-rich banks. The thick fringe of reeds and other flowering plants provides homes and food for water voles and others. Where the cattle cross the river the flora changes.

muddy waters, and the muddy bank provides the ideal place for the extensive tunnels of a water vole's burrow. The young are born naked, blind and helpless, in a chamber of the burrow lined with grasses and rushes.

The dipper feeds on the bottom of upland streams and finds a place to build its mossy nest on overhanging trees or ledges of bedrock, often near a waterfall. The kingfisher has a territorial perch on a branch over the river, but requires a sandy bank to excavate, with its strong beak, a tunnel more than half a metre long, ending in its nest hole (p. 89). There are other competitors for vertical sandy banks such as colonial sand martins which excavate equally deep tunnels. The long tunnels ensure the safety of the young, deep inside the eroding sandy banks. The river provides abundant emerging insects on which the sand martins feed their young.

Shingle islands are often isolated from terrestrial predators by deep channels of water and so make suitable breeding areas for the ground-nesting birds such as swans, ducks, moorhens and coot. These birds choose the well vegetated islands that are more stable than the recent deposits of shingle, and will often nest beneath clumps of reed canary grass or hemlock water dropwort, which forms a dense cover over them. Swans, however, will often nest much more in the open because they are able to defend their nest site aggressively.

In well wooded valleys the bank deposits may vary from boulders to mud held together by the roots of waterside trees. On these muddy banks there may be traces of otters (p. 88), but the otter holt is well hidden within the roots of a waterside tree, where it is well protected from erosion. Here, the blind, helpless young are born. But otters that live in the marshy estuaries where there are extensive reed beds rear their young in a nest among the reeds lined with their purple flowers.

On some of the tree-shaded muddy banks the liverworts form a close green carpet. This is a group of plants that are dependent on the damp or humid air close to the river in order to complete their life cycle. These primitive plants have not developed the specialized tissues associated with stems, roots and leaves, so are restricted to creeping across bare surfaces in damp, shady places where little else can grow. Their method of reproduction further limits them to wet habitats as the male spermatozoid has to swim to the female archegonium. After fertilization, a capsule develops and grows up from the parent plant on a long translucent stalk. When conditions are favourable in early spring, these capsules are quite abundant. On a dry day, they open to release the spores. But dry days may be rare, and *Marchantia polymorpha*, a common liverwort along stream edges, can also reproduce itself in wet weather. It has special 'splash cups' in which gemmae (packets of cells produced vegetatively) are dispersed from the parent plant with rain drops. Ferns are also common along shaded stream edges. Sexual reproduction is much like that of the liverworts, but occurs on a tiny, heart-shaped prothallus quite distinct from the familiar fern plant which produces the spores that give rise to the prothalli. Ferns, like mosses, therefore have a distinct alternation of generations, whereby the spore-producing plant alternates with the gamete-producing thallus.

Aliens and invaders

The river's edge provides an ideal habitat for invaders, as each year new areas are added or exposed which allow rapid colonization by new arrivals. The willowherbs are common along the stream edges, and rosebay willowherb frequently takes the place of its relative, the hairy willowherb. As each plant can produce on average 80,000 wind-blown seeds that will colonize any newly cleared ground, its invading potential is enormous. Formerly restricted to hill regions, it has greatly increased its range as a result of man's activities in forest clearings, road edges and other places.

The yellow monkey flower is a North American plant which was introduced to Britain in 1830, and has now spread to many British and north-east European stream banks, where its handsome, red-spotted yellow flowers are conspicuous in July and August. The seeds are small and borne in a capsule, containing around 150 seeds, which are dispersed by the water or on the feet and feathers of water birds (pp. 20–1). The exploding capsules of policeman's helmet, or Himalayan balsam (p. 83), have shot the plant around river banks in Europe since it was introduced in 1839 from the Himalayas.

The American mink has been bred for its fur in Europe since 1929. It was reported to be breeding in the wild in 1956 and has since established itself along many streams and rivers, where it competes with the otter for food and breeding sites. Another introduced mammal which has escaped and established itself in the wild is the South American coypu. It feeds and breeds in reedy areas such as the Norfolk Broads in East Anglia. But the native beaver, formerly widespread in forested wetlands, has disappeared from Britain, having been hunted almost out of existence by the thirteenth century, and is now much restricted in Europe (pp. 88–9).

PROJECTS

1 Find a river bank to which you have easy access and can visit regularly. Make sure the bank is safe and not in danger of collapsing. Record the slope (p. 15), water level, aspect, bank substance, and management of the adjacent land. Make a belt transect 1 metre wide and mark its position with two canes well above water level, so that you can return to the same spot each month. Record the bank plants, whether they are flowering or fruiting, and their percentage cover in your rectangle. How does the cover change throughout the year? What other factors change over the winter and following spring?

2 Visit a number of streams and rivers on a variety of substrates, and record the plants on banks of different substrates, heights and so on. Do different plants require different bank conditions?

A water vole *Arvicola amphibius* swimming among the fringe of reeds on which it feeds at the edge of a muddy river bank.

3 Walk along a length of river bank to which you have easy access and record any signs of animals using the bank (see pp. 88–9) and plants that are grazed either from the land or from the water.

Water vole holes in a river bank.

RIVER MANAGEMENT

Rivers have long been used for transportation and fishing, and are now also used as a source of water for hydro-electric power, irrigation and recreation.

Operations to provide flood prevention and control are necessary to avoid damage and deprivation to the people living along the river banks, particularly as the built-up areas enlarge; this results in a more rapid run-off of storm water. To meet the many needs of man, river management is essential, particularly as there can be conflicting demands on the river's resources.

Anti-erosion measures

Flood prevention often leads to the canalization of rivers. After the banks are straightened, they are reinforced to stop erosion, and submerged aquatic vegetation is cut or dredged out. This can result in local stretches of river that have little visual appeal and may not provide birds and mammals with ideal habitats in which to breed. Fishermen, however, prefer to have weed-free areas for casting their lines, and the overhangs are cut away so fish cannot retreat beneath them.

Bank stabilization is usually carried out where slumping is liable to occur or where the banks are continually used by the public. The most effective method in areas of strong scouring is the building of concrete or metal-sheet retaining walls, both of which are ugly, and inhospitable to wildlife. Drystone walls or sacks filled with clay, sand or concrete are used along short lengths of river, but if extensive bank stabilization is needed, brushwood borders which trap silt, or reed and brushwood mattresses, are surprisingly effective. In East Anglia, the water authority grows osiers especially for weaving into 9 × 6-metre mattresses which are towed by boats to sites where the river is scoured by tidal action, and then sunk with stones. Some of these mattresses have been in place for forty years and still show no signs of needing to be replaced.

Making amends

In Russia, turbines on the river Volga prevent the upward passage of adult sturgeon, which have to be caught and stripped of their eggs; these are then artificially fertilized and reared in hatcheries so that stocks are

▲Fly fishermen electro-fishing a river in winter to remove the coarse fish in a stretch which has been stocked with trout.

◀Watercress beds thrive in winter in the clear waters diverted from a chalk stream.

▼An automatic 24-hour water sampler in use beside a river where it can take twelve samples during a 24-hour period.

▲Reinforcement of a river with metal sheeting to prevent further erosion of the bank.

▲Cutting submerged vegetation in the river Test, in the south of England, in July, helps the water to flow and provides clear water for the fly fishermen to cast their flies.

▶Hatchery-reared trout rising to food in a stew pond beside a river, where they are contained until they reach a size for re-stocking the river itself.

maintained. Similarly, in Sweden, the extensive use of the rivers for floating timber downstream, plus the numerous hydro-electric schemes, means that the salmon stock has to be maintained by stripping the adults at the beginning of their attempt to migrate up to the headwaters, and rearing the eggs through to the smolt stage. Many power stations use vast amounts of cooling water from rivers which is then returned at a higher temperature. On the river Danube the cooling water has raised the river temperature by as much as 10°C along some stretches. An increase in temperature reduces the oxygen content sufficiently to stop the spawning migrations of a number of fish that are confined to the Danube system, and so threatens them with extinction.

The discharged warm cooling water need not always have such catastrophic effects. In some places it is utilized by fish farms which oxygenate the warm water and use it for the rapid growth of certain fish. Carp, eels and trout are favourite species for culturing in this way by several countries including Russia, Poland, France, the Netherlands and Britain. The grass carp is a vegetarian species which is cheap and easy to feed and has been introduced from China to a number of countries for such fish farms. It is not yet being farmed on a commercial scale in Britain (pp. 174–5).

The water from cooling towers of power stations can actually improve the water quality in polluted rivers with a very low oxygen content since it can help to dilute the pollutants. In 1970, when some sewage works in Britain were affected by strikes, cooling towers were brought into action to help save the fish populations of rivers in which untreated effluents were discharged.

When the Pitlochry dam was built on the Scottish river Tay—one of the most valuable salmon rivers in Britain—a 311-metre-long fish ladder with 34 pools was constructed. Over this stretch of river, the salmon cover a 15-metre rise by moving through the pools via interconnecting pipes. Large glass windows have been built along the sides of one pool for visitors to watch the migrating fish.

Monitoring the water

The most important aspect of river management is the careful control of the water quality, which is carried out by repeatedly sampling the water and testing for impurities, and, when necessary, regulating the volume and quality of discharges into the rivers. All pollutants which pass into river systems in small concentrations are soon diluted on their passage downstream. However, chemicals and synthetic materials are new pollutants which, if they are not biodegradable, can cause havoc to a river system. The hardest pollution to control is the sudden introduction of high concentrations of toxins such as cyanide, through accident, incompetence or sheer irresponsibility. The poisons can kill and then disappear out to sea before they are identified and the source confirmed. Fortunately, the unsightly clouds of non-biodegradable detergent foam on waterways are now largely a thing of the past. Successful control of water quality is vital for both environmental health and our own health. If the nitrate content of drinking water were to become too high, young babies would suffer brain damage, and if the treatment of sewage became inadequate, a whole range of water-borne diseases, such as typhoid, paratyphoid, cholera, amoebic dysentery, infectious hepatitis, poliomyelitis and salmonellosis, would become rampant. When an epidemic of Asian cholera caused the deaths of 50,000 people in England in 1931, it was not known that water could transmit the disease.

Fishery management

Regulation of fishing is necessary to maintain both the quality and the quantity of a river's fish stocks. The fish stocks may be managed by removing unwanted species from stretches of water by electro-fishing. If an alternating current is used it stuns the fish, and they can then be removed by handnetting; if a direct current is used the fish are drawn towards one electrode and they must be caught before they touch it, otherwise the electric current will kill them. Reaches which have had coarse fish, such as pike and perch, removed by electro-fishing can then be re-stocked with rainbow or brown trout reared in hatcheries.

Hatcheries are usually sited where there is a copious supply of cool, clean water, often using springs as their source. Although fish grow faster in warm water, the incidence of fungal and bacterial diseases also increases as does the susceptibility of the fish to pollutants. By isolating the stock from possible sources of infection upstream, the chances of introducing disease are greatly reduced. A high flow is also desirable so that any uneaten food and the young fry's faeces are swept away and do not add to the oxygen demand of the water.

On chalk streams such as the river Test, in the south of England, renowned the world over for its trout fishing, the submerged weed growth is so prolific in summer that it raises the water to a higher level than during the winter when more water is carried in the river. The submerged weeds are therefore cut throughout the summer both to prevent flooding and to provide clear areas for the fly fishermen to cast. Almost the entire length of the Test is highly managed for fly-fishing: the weeds are cut; the coarse fish are removed by electro-fishing; coot and moorhen are shot; the bankside vegetation is cut down for easy access to the river's edge; and casting platforms are built. It is still, none the less, a very beautiful part of the Hampshire scenery.

Watercress culture

The crispy green shoots of watercress, which have become so popular in recent years, originate as a crop grown in beds of slow-moving water adjacent to chalk streams. England and Wales alone grow about 127 hectares. Each bed is cropped about five times a year. In April and May the beds are cleared of the old plants and reseeded; cropping commences six weeks later. Freshwater shrimps abound in the beds and, although they cause no problem themselves, they attract wild duck which move in to feed on them. As the duck feed, the plants are uprooted, which can cause quite extensive damage. The beds are also periodically infected with a fungal disease called crook rot, which can be treated with a zinc compound. Application of the compound needs to be timed so that it does not harm downstream inhabitants.

Considerably more research into monitoring techniques is required so that the best use is made of rivers and the conflicting demands made upon them are reconciled.

RIVERS IN THE LANDSCAPE

Although the river bed and banks change constantly, the drainage pattern that a river cuts in the landscape is characteristic of the rock type and rainfall of the region.

Rivers are probably the oldest features of our landscape, since it is by their action on it that the present contours and valleys were created. Once they begin to cut a valley they are trapped in it, and, unless there are great changes in rainfall or in the tilt of the earth's crust, they will continue to cut the same valley for thousands of years. Each river is dependent on the many tributary streams that feed it along its length, these being fed by rain that arrives in their catchment areas. The streams may receive the water either as it runs off the surface of the land or from the underground reservoir from which it emerges at ground level. The amount of water that the ground can hold is known as the 'storage capacity', which increases with the permeability of the rock. The rainfall and rock type will determine not only the pattern of the river system, but also the plant and animal species living in it and the distribution of the settlements on its banks.

In highland regions the rock is often hard and impermeable and the soil thin, so the rivers are fed by surface run-off water, which varies with the amount of rain falling. These rivers are susceptible to sudden changes in water level and flow, and in turbidity caused by the sediment that the water carries. They have great erosive potential, and tend to cut steep-sided valleys towards the main river. This dendritic, or tree-like, pattern is repeated in all the valleys on this rock type and is obvious along the resistant rock areas in the west of Britain and France, and in the regions of the Alps and the Pyrenees. Even when these rivers cross lowlands to the sea, they still behave like upland rivers, with all the impetus that the gradient gives, and have extensive shingle banks and little fringe or submerged vegetation until they reach more extended lowlands or estuaries. Here man is wary of settling too close to the

▲The river Asseca in Portugal is on semi-permeable sandstone and is fed from the ground water table. There are no traces of storm flows and the water supports a rich fringe of monocoty-ledons and oleander shrubs *Nerium oleander* as well as submerged plants and blanket weed.

▲This river, in the Ordesa National Park in the Spanish Pyrenees, is fed by run-off water which carries large amounts of material eroded from steep valley sides.

▼A dry river bed of the Nedd Fechan river in south Wales whose waters have been captured by the river Neath.

▲Gordale Scar in Yorkshire in the north of England has been created by the action of water on hard carboniferous limestone dissolving an underground route of swallow holes and caves through the rock. These have collapsed to leave a vertical-sided limestone gorge.

▲A reconstruction of the old mouths of the Rhine.

▼This river and its tributaries rise on chalk downs. With this stable water level and flow, the settlement clusters around the river, and the lime-rich water supports a diversity of plant and animal life (p. 69).

river, and villages are usually on higher ground out of reach of a storm flow.

Where rivers run on permeable rock, such as chalk, soft limestone or sandstone, the villages nestle close to the river with an air of ancient peace and tranquillity. This is because, even at times of high rainfall, the river does not rise rapidly, much of the water being held in the ground and only slowly released into the river. Rolling farmland often continues to the edge of the river. The water is clear and there is plenty of submerged and fringe vegetation characteristic of each rock type. On soft sandstone, as in the Algarve in south Portugal, there are fringing bands of monocotyledonous plants and shrubby oleanders as well as a few submerged species. On chalk and soft limestone, the water is clearer and the fringe richer, with watercress, lesser water parsnip, brooklime and many others, while the bed of the river sparkles with clumps of water starwort, water crowfoots and pondweeds (p. 69). The water acts very differently on this landscape, dissolving the calcium carbonate, of which the rock is made, with carbonic acid contained in the surface waters that have been in contact with the carbon dioxide in the air. Hence the rivers carry little sediment and the water is very hard, being full of calcium bicarbonate in solution. The rivers themselves are found only in the lower valleys, yet there are many dry valleys on the hills around, where either solution weathering has gone on below the surface or the water table has dropped since the valleys were formed.

Limestone gorges

Rivers running over hard limestone produce a particular landscape of steep-sided gorges, swallow holes, caves and underground rivers as in the Dordogne in south-west France or in the Pennines in the north of England. Although the rock is dissolved by the action of water on its surface, it is very hard and not easily eroded by water action. It is also not very permeable, so the storage capacity is not large and these mountain limestone rivers are often subject to storm flows that remove all fringe vegetation. However, submerged vegetation often survives to dominate the river in the next growing season.

River patterns

Water always cuts the easiest route downwards, so where the rock type changes, or where the sediments of which the rock is made are layered in bands of hard and soft rock, the river pattern will change, following the direction of the softer rock and producing a characteristic pattern in that region. But erosion does not only proceed downwards from the source of

a river. It may also cut back at the source, so lengthening the valley. If by this process it cuts through into another river system, it may capture the water of that river, changing the flow, so that, eventually, only a dry river bed may remain from the earlier route. In time this will be filled in with soil, which will cover the evidence that remains in the form of rounded, water-worn pebbles, until an investigator looks below the surface.

It is from evidence like this that we can reconstruct changes that have happened in our river systems over the last million years. To the landward side of the present coastline of East Anglia there are river deposits containing stones of slate, chert and grits from the Ardennes and parts of western France and Belgium. This deposit was laid down by the river Rhine around one million years ago, when Britain was still joined to the European mainland. On the bottom of the North Sea lies the evidence of its routes, and the land that the river passed through. The original fen forest gave way to Arctic flora during the Ice Age and then to birch and willow wood before the sea inundated the land, leaving the mouth of the Rhine at Rotterdam as it is now. This rise in sea level that accompanied the melting ice drowned many of the river valleys on the western coasts of Britain and France. These now have extensive tidal waters and are infilled with recent sediments, leaving the original valley bottom metres below. There are no historic records for these changes; they are buried in or carved out of the land itself.

There are other changes to the earth's surface that may leave their mark on a river system: the tilting of the earth's crust, or the intrusion of new material, such as lava, may change the direction of water flow. As recently as 1783 the largest lava flow in historic times blocked the Skaftá river in Iceland, so that the river had to cut another route around the lava.

In human terms rivers are ancient features, being the sites for early settlements, such as the caves of the Dordogne, which were occupied by paleolithic hunters around 26,000 years ago. Later settlers used rivers for transport from one settlement to another and for power to drive flour mills and other machinery. But in all this time the same power that drove the mills has caused great disasters when there was too much water. Today water storage and flood prevention are important and expensive items in our society, and will depend upon accurate data for rainfall and storage capacity. When the alterations necessary to carry out these measures are made, the habitats and river pattern may also change.

PROJECTS

1 Using a large-scale map, of 1:100,000 if possible, trace off the rivers in your region. Can you detect different rock types from this drainage map? With the aid of a geological map, repeat the tracing in a variety of rock types. If you can, visit these rivers and build up a record of their banks, vegetation (fringe and submerged) and depth (many water authorities have depth scales at various sites along a large river). How do these vary?

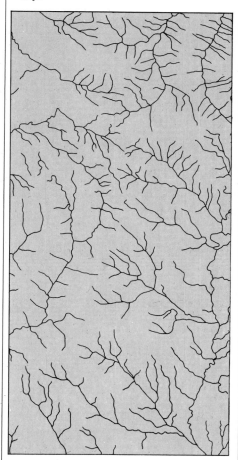

A river drainage pattern on resistant rock.

2 Where two rivers from different geological regions join, investigate the flora and fauna of each river above the confluence. Does the habitat change below the confluence, and which species survive in the river below?

3 Collect water from different types of river, and at different seasons, by lowering a litre bottle into the river from a bridge to determine how much suspended sediment the water is carrying. Pour the water through filter paper to collect the sediment. Feel the sediment to estimate particle size (silky = silt, rough = sand). Test the pH of the filtered water, using pH papers. Do the sediment type and amount vary with the rock type?

A CHANGING RIVER

The Thames used to be one of Europe's filthiest industrialized rivers. Its transformation from a stinking sewer into a river inhabited by many fish and birds gives us hope that man can remedy the pollution of waterways elsewhere.

The River Thames around London from Hampton Court to Gravesend

Kew Bridge
Richmond
Teddington
Hampton Court
Fulham
Westminster
London Bridge
Southwark Docks
City of London
Docks
Sewage Works
Woolwich Reach

On its way to central London, the Thames snakes north-eastwards around Hampton Court Park and its Palace past the Royal Botanic Gardens at Kew. Henry VIII regularly used the Thames to commute from London to Hampton Court Palace. Many of the riverside mansions and landscaped gardens which were a feature of the Thames as he knew it no longer survive, and the banks have become much more built up.

The past
Since Roman times, the Thames has provided a thoroughfare from the North Sea into the heart of London, which, by the nineteenth century, had become the largest port in the world.

People lived and worked on the Thames, which also provided a valuable source of food. Eels were numerous, and were caught in sieves and buckets, while even trout and salmon were so abundant that they were eaten by the poor. Along the tidal stretches, smelt bones have been found in layers dating back to Roman times at Southwark, and to the Middle Ages at Westminster, and remains of sturgeon have been unearthed in medieval remains from Westminster Abbey.

Many uses of the river were conflicting; during the Middle Ages, monasteries and mills utilizing water power were built beside the river. Weirs were built to catch the fish and to power the mills, but these impeded the passages of the barges. Some movable sections, known as flash locks, were later built into the weirs, but barges still had to wait until the main rush of water had subsided before they could safely pass. The first of the modern pound locks with two sets of gates was built on the Thames in 1630.

In times gone by, the river froze over, allowing Frost Fairs to be held; the last was held in 1814. The tidal Thames, which extends from the mouth of the estuary to Teddington Weir, no longer

▼Aerial view of the Thames Barrier under construction, as at July 1980. From bottom left: the main control tower, generating building and workshop building still incomplete; four completed piers (the fourth lacking only its stainless steel shell); two main piers and three smaller ones still under construction. Downstream is to the right.

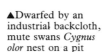

▲Dwarfed by an industrial backcloth, mute swans *Cygnus olor* nest on a pit beside West Thurrock Power Station, adjacent to the Thames.

▲Driftwood on the Thames is a serious problem to powered and sailing craft, as well as being unsightly and unhealthy.

▲Swan-uppers in their skiffs on the Thames in July 1981. The ancient and picturesque ceremony of swan-upping takes place each summer on the Thames, when mute swan cygnets are caught and bill-marked for the Livery Companies of the City of London. Cygnets from unmarked swans belong to the Crown.

▲A composite picture, illustrating the scene at Westminster if the Thames should seriously flood, was used for advertising purposes by a company which provided components for a temporary flood barrier.

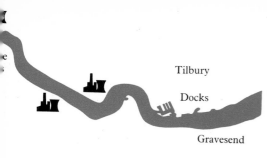

Tilbury

Docks

Gravesend

Power stations at which fish are caught

5 km

freezes because of the increased flow rate produced by the deepening of its channel. In the severe 1962–3 winter, however, it froze over above Teddington.

For centuries, the Thames, like other rivers, was treated as a rubbish tip, into which all wastes were thrown.

The introduction of the water closet led to the discharge of untreated sewage waste into the Thames, making it an open sewer. By the middle of the last century, the stench was so overpowering that sheets soaked in disinfectant had to be hung outside the Houses of Parliament at Westminster. Several cholera epidemics broke out. Sewage works were built, but they soon proved to be inadequate to cope with the increased sewage and industrial effluents. Shipbuilding had also thrived in the nineteenth century, and, as the docks extended on the lower flood plain of the river, the wetland sites disappeared.

The present
The river's highly looping course from Teddington to Gravesend causes the main tidal stream to switch continually from bank to bank, thoroughly mixing the fresh and sea water so that there is a gradual salinity decrease upstream (pp. 102–3).

The deterioration in quality of the river water led not only to the decline of fish, but also to that of the birds which fed on them. As recently as the 1960s, few birds occurred along the inner Thames (the 40 km stretch between London Bridge and Tilbury). Today, this stretch is home for large numbers of overwintering wildfowl (10,000) and waders (12,000). In 1957 the 64 km stretch between Richmond and Tilbury contained no fish, apart from eels, which could surface to breathe. Fish began to return six years later, and in 1965 live specimens were caught on the screens of the cooling water intake of Fulham power station. Regular surveys of fish caught on the intake screens of

eight power stations provided proof that more and more fish were returning until, by the end of 1975, 86 species of freshwater and marine fish had been caught. How has this dramatic reversal taken place?

In 1959, an anti-pollution campaign was activated by the Port of London Authority and the Greater London Council. The long-overdue clean-up programme was only possible when the special problems of the Thames were fully understood, such as the long retention period of water (including pollutants) in the tidal Thames despite the twice daily ebb and flow of the tide (pp. 100–1). A stick dropped into the river at London Bridge will take at least 20 days to float the 64 km out to sea, and may take as many as 80 days if the flow of fresh water is low after a period of dry weather. The sewage works were improved by installing better filtration, treatment and aeration plants so that the discharge water is now almost pure. The sewage works can now handle the normal value of domestic and industrial waste, but still cannot cope with large amounts of surface run-off water during freak storms. Then, untreated sewage may be discharged into the river through the storm sewers, resulting in a rapid drop in the oxygen content of the water and causing fish mortalities.

Stringent bye-laws now control the discharge of pollutants by riverside industry, and this has also helped to stop the depletion of the oxygen content of the water which is constantly monitored.

The river is no longer a health hazard, but modern development still presents problems. Every year, 1,500 tonnes of rubbish are collected from the river.

The most dangerous forms of rubbish are plastic (especially plastic packaging from building products), rope, timber and paper bags, which foul boat propellers. It is an offence to dump objects likely to be a hazard to navigation. Even so, baulks of timber from old decaying warehouses and jetties still float downstream, and have to be collected from the water.

Another hazard which has been highlighted in recent years is the poisoning of swans by anglers' lead shot. This factor, probably more than any other, was the cause of the catastrophic decline of the Thames mute swans in the late 1970s. Each year, during a week in July, cygnets on a non-tidal stretch of the upper Thames are caught and marked during the ancient swan-upping ceremony. Six boats (three pairs, carrying swan-uppers representing the Crown and two City Livery Companies, the Dyers' and the Vintners', and each sporting its own banner), make their way up the Thames until they find a family of cygnets. The

boats encircle the cygnets, trapping them against the bank. The bills of parent birds are examined for marks: Vintners' birds have two marks on the upper bill and Dyers' birds one mark on the lower bill, while the Queen's birds are unmarked. The cygnets are given the same marks as the parent birds. Until recently, all the cygnets were pinioned (their flight feathers cut) to ensure that they remained within the same stretch of the Thames; but now pressure bodies have brought about the abandonment of this practice. Consequently, marked birds are free to fly elsewhere, so this ancient custom may eventually disappear.

The future
The disastrous 1953 floods along the east coast of Britain and the Thames estuary, which cost 300 human lives, highlighted the real danger of catastrophic flooding of the tidal Thames. Floods in London are not new. In 1236, the Thames overflowed into the Palace of Westminster, and men were able to row along the Central Hall. Each year the risk of flooding increases, partly because the bed of clay on which the city is built is slowly sinking, and partly because high tide levels are rising; at London Bridge they have risen 60 cm during the last century. Today, there would be a real danger of catastrophic flooding in central London if a high spring tide coincided with a storm surge and a high river outflow. The flood walls have been built up as high as they can be without spoiling the character of the river.

After years of research, the decision was made to build a rising sector gate barrier across the Thames in the Woolwich Reach. The movable barrier, which will be the biggest in the world, is due to be completed in 1982. It will hold back surge tides passing down from the upper estuary and thereby safeguard the lives, homes and jobs of more than a million people who live and work in central London. The barrier consists of a series of movable gates, each of which is pivoted. Normally, these gates will rest in the river bed, but within 30 minutes they can be swung through 90 degrees into a vertical position.

The wildlife of the Thames would not be threatened by raising the barrier for short periods when the danger of flooding is imminent. If the barrier were to remain permanently closed, however, it would prevent migratory fish from moving up and down river; the silt washed down from the upper reaches would accumulate and have to be dredged out (pp. 176–7); and the warm water discharge from power stations would reduce the quality of river life by heating up the water (pp. 170–1).

CHANNELLED WATER

From early times man has channelled water for his own ends, yet with time, plant and animal life fills every niche of the managed habitat.

Ever since man moved into permanent dwellings, tilled the land and kept domestic animals, he has channelled water. Once he had built settlements he made drains. Then his fields needed drainage or irrigation channels, and by Roman times he had discovered that canals were a convenient way of carrying heavy loads across flat country or from river to river.

After the invention of locks, barges could even go uphill, and by 1681 the Bay of Biscay was connected to the Mediterranean by the Canal du Midi, with 119 locks to take traffic 209 metres above sea level. With the birth of industry, canals played an essential role in transporting materials and goods from place to place, and by the middle of the nineteenth century there were continuous networks of canals and rivers across Britain, mainland Europe, and America. This meant that water plants and animals also had a continuous route around the countryside; a Thames snail could travel from London to north Yorkshire, to the Midlands, and through the Avon to Hereford, if it could withstand the pollution on the way.

Water that is channelled offers particular conditions for animals and plants, brought about by the construction, use and maintenance of the channel. Drainage and sewage pipes are constructed from a smooth material so that cleaning can be carried out by flushing with more water. There are few anchorage sites here for animals and plants, although some species of bryozoa encrust the interior of water pipes. The pipes must be able to carry a considerable amount of water, as in city areas all rain landing on impermeable concrete and tarmac will end up in the drains. Irrigation and land drainage channels are more difficult to keep clear of plants and animals, and must be regularly cleared every year or two of all the vegetation and wildlife that invades them. The gradient that keeps the water flowing is critical, and in many areas of low-lying Britain, Holland and Germany the water must be pumped into the sea to maintain the flow, particularly at high tide. The banks of

▲Digging out a ditch with a mechanical digger benefits the floating plants by removing the fringe emergents.

▶A man-made dyke, dug to drain wetlands. The common reed *Phragmites australis* is rooted at the edge of the dyke, but where it is cleaned out the water violet *Hottonia palustris* flourishes. The delicate green rosettes are below the surface throughout the year, but in June the flowers appear above the surface.

▼An outfall sewer at Charlton, south London. The Thames Water Authority treats and disposes of 43 million hectolitres of sewage each day.

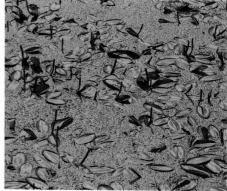

▲This drainage channel has little flow of water, so that free-floating duckweed *Lemna* spp is abundant around the leaves of the bottom-rooted pondweed *Potamogeton natans*, whose flowering spikes are visible above the surface.

drainage channels and ditches must be straight, and cut deep enough to allow the water to drain out of the surrounding land. A canal, on the other hand, must be water-tight, especially where the water is maintained on a slope or across a valley. The banks must be strong enough to resist the wave action caused by the boats, a problem which has increased recently with the replacement of the old horse-drawn boats by engine-powered ones. The water-tight base used to be made of puddled clay, but now it is frequently concrete, thus reducing the living sites available for animals and plants. Boats using a canal generally keep it clear of larger plants, but as soon as a canal is abandoned the reedswamp vegetation moves in.

Conditions for life

Most channelled water is highly nutrient-rich (eutrophic) but oxygen-deficient. In the early days canals were often open sewers in city areas, with clouds of methane gas above them, caused by anaerobic respiration of bacteria, that occasionally caught light. Today they are cleaner, but are still nutrient-rich from run-off water and from the organic matter produced by the community that inhabits this environment, which is stable, because water levels are maintained and the slow rate of flow has no erosive power. However, any animal or plant living here must be tolerant of the conditions imposed by boats. Frequent wave action and turbid water lower the number and variety of submerged plants and of fringe emergents, which in turn restrict the species and numbers of herbivores that feed on them. Where boat traffic is restricted, many wetland species, including various coarse fish, are found in the canals, so at a weekend fishermen will be sitting every few metres waiting for their floats to bob, with a perch, roach or tench (p. 44–5). A dip with a pond net in one of these canals will produce many invertebrates familiar from other still water habitats, such as the water boatman, great diving beetle, leeches, flatworms and abundant snails (p. 32). But it may also bring up some of the food plants that are below the surface, such as the beautiful ivy-leaved duckweed, the hornwort, species of pondweed and the ubiquitous Canadian pondweed, whose route into Europe was by the canals (pp. 20–1). Other species have found this habitat congenial, and there is little doubt that boats have aided the spread of species like the sedentary branched colonies of the brackish water hydroid *Cordylophora lacustris*, which first appeared on wood and ships' hulls in the canal locks in Dublin and Belgium in the 1850s, then spread into France and Germany and over lowland Britain. The zebra mussel is another brackish water species that has invaded far inland through rivers and canals since the eighteenth century. The freshwater sponge is often found in greyish-green masses at the edges of canals (p. 68).

There are other species, however, that are now rare in habitats other than canals and drainage ditches, as if these have provided the last bastion of a type of wetland that has been destroyed elsewhere. In a few areas of midland England and mainland Europe may be found the delicate floating water plantain *Luronium natans*, or the starry green rosettes of water violet, which support a spike of pale purple flowers. Floating plants abound on these slow-flowing waters, and actually benefit from the regular cleaning of the ditches as this prevents fringe plants from shading them out. Conditions in the ditches, or in the less disturbed and oldest canals, aid the survival of some of the most beautiful water plants. Ancient water courses, such as the irrigation channels for the rice paddies in the Po Valley in Italy, have growing in them a great variety of submerged plants which tolerate the varying water levels. Another species that has survived in this habitat is the rare great silver water beetle, which builds for its young a floating silken nest with a snorkel-like spur that appears above the surface.

Although these richer canals and ditches are a far cry from the sewers of our cities, where little can survive, there is plenty of evidence that species have adapted to changes made by man during the last 300 years and that today man-made canals and channels are a rich reservoir of plant and animal life.

PROJECTS

1 Find a man-made channel or drainage ditch. Where does the water come from? Is there any farmyard or other effluent entering the water? How is it managed, and how frequently? Sample the water and the substrate for plant and animal life. Identify the species or genera and find out how long the life cycle is. Does this fit into the management routine? Do the species that you have found indicate any form of pollution (see p. 84)?

2 Make a plant survey along the edge of a canal, recording substrate and water depth for each species growing in it. Does the type of bank and the frequency of boat traffic affect the species growing there?

The Grand Union Canal, built to supply midland industry in Britain, now carries holiday-makers in motor-powered longboats or launches. The wash from these boats and the turbidity caused by the propellers restrict fringe and submerged vegetation.

3 In an area where dykes are regularly cleared, can you work out the order in which plant species appear after clearance? Record each species. Then find a dyke that has not been cleared for some time and record each plant species that you find in it. Which has more species living in it? By examining dykes which have differing clearing routines, can you estimate which species are tolerant of each routine?

THE EDGE OF THE SEA

The interactions which take place where sea meets land can bring about profound and rapid changes in the coastal geography, which result in an especially dynamic habitat.

The one factor which dominates natural, as well as human, life on the coast is the relentless tidal rhythm which moves sea water up and down the shore, day in and day out.

Tides

Most European ports experience a semi-diurnal tide, which means that there are two high and two low tides during each 24-hour period. The tides form as a result of the gravitational pull on the oceans by the moon. The part of the earth nearest to the moon will experience a high tide as the water is sucked towards the moon in a high tidal 'bulge'. The other side of the earth directly opposite will also experience a high tide, and, to compensate for these high tides, other parts of the earth's oceans will then have low tides. As the moon rotates around the earth, so the high and low tidal areas are constantly changing.

Although the earth rotates once on its own axis every 24 hours, it takes another 50 minutes for the earth and the moon to return to their same relative positions so that the actual times of low and high water are not constant each day, but become progressively later.

The vertical height between high and low water (the range) also varies daily. During the new and full phases of the moon, when the moon and sun lie in line with each other, their combined pulls result in the high-ranging 'spring' tides. These tides are therefore not associated with the season of spring, but occur throughout the year, alternating with the small-ranging 'neap' tides, which occur during the first and last quarter phases of the moon, when the moon and sun lie at right angles to one another and so their pulls work against each other.

Very spectacular high-ranging spring tides occur twice a year at the time of the spring and autumn equinoxes, around 21 March and 21 September. At these times, when the tides penetrate higher up the shore and also recede further down the beach, shore excursions are especially exciting, since the upper fringe of the sublittoral oarweed forests become exposed to

▲ A creek winds its way through the estuarine muds of the tidal part of the Beaulieu river, in the south of England, exposed during low tide. Within the sheltered confines of the estuary, the sea has lost all its destructive power on the land, and it merely mixes with the fresh water to produce a brackish water habitat.

◄ The eroding power of the sea has carved this conglomerate and basalt arch at the Reykjanes peninsula in Iceland.

▼ Coastline profile showing how a cliff, a notch, a beach and a wave-cut platform are formed by a combination of erosion and deposition.

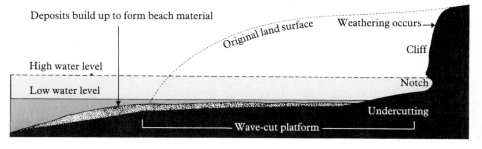

Deposits build up to form beach material

Original land surface

Weathering occurs →

Cliff

High water level

Low water level

Notch

Undercutting

Wave-cut platform

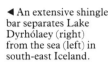
◄ During winter storms, the sandy floor of this beach in south Wales, which is exposed to the full force of the south-westerly gales, may be completely scoured out, but by the following summer, the sand will have gradually accumulated again.

◄ An extensive shingle bar separates Lake Dyrhólaey (right) from the sea (left) in south-east Iceland.

▼ An atypically calm sea laps against the edge of a wave-cut platform exposed during low tide. More typically, this part of a beach in south Wales is battered by 10-metre-high waves crashing on to the platform.

landlubbers (pp. 134–5).

The times and ranges of high and low tides are therefore predictable years in advance, which greatly aids the planning of commercial shipping routes and times, and, of course, biological research. Anyone who wishes to study life in the intertidal region should be equipped with local tide tables, which are usually available from coastal fishing tackle shops or newsagents. In most places, it takes about six hours for the tide to move up (to flood) the shore from low-water level to high-water level and the same time for it to fall (to ebb)

back down again. In restricted tidal inlets, as in the opening of some fjords or sea-water lochs, powerful tide rips can develop which produce highly oxygenated water. If this coincides with a plentiful food supply, the bottom will be richly carpeted with massed concentrations of brittle stars, sea anemones, sea cucumbers and sea squirts.

Land lost to the sea
If the tides rose and fell with a flat sea level, the action of the sea on the land would be minimal, as can be seen in

enclosed sea-water lochs; but we all know that the surface of the open sea rarely resembles a mill pond. It is the power of the waves which constantly attacks rocks and moves sand and shingle. Wave height relates to both the wind strength and the extent of open water, or fetch, over which the wind blows. When a wave reaches a shore, the crest surges forward carrying with it shells, sand, shingle or any other debris. The water which surges up the shore is known as the 'swash', and after the wave has broken, the water falls back as the 'backwash'.

When the height of waves is exaggerated by very strong winds coupled with spring tides, their destructive force can be catastrophic, as occurred in the North Sea on the night of 31 January 1953. Winds of up to 160 km per hour coincided with spring tides and low atmospheric pressure, resulting in a storm surge which caused serious flooding in Britain and on the west coast of mainland Europe. In Holland alone 18,000 people died, 50,000 houses were destroyed or severely damaged and 162,000 hectares were flooded.

One of the initial stages in the marine erosion of a landform is the formation of a wave-cut notch where the sea undercuts the land. Further erosion results in a wave-cut platform gently sloping seawards, much of which will be exposed at low tide. This platform may remain as bare rock or get covered with a shingle deposit. Hard rocks are better able to withstand marine erosion than soft alluvial cliffs, but any land which has to face the brunt of pounding waves is susceptible to erosion, and when waves hurl pebbles at cliffs, they will accelerate their cutting back. Soft rocks are eroded rapidly so bays are formed, while the hard resistant rocks remain jutting out as headlands (pp. 154–5).

New shores from old
At the same time as land is being eroded away at one place, it is being built up elsewhere. Sandy bars, beaches, coves and tombolos, as well as shingle bars, ridges and spits, all arise from the deposition of eroded material by the sea. Even when a sandy beach is formed, it is not necessarily a permanent coastal feature. Summer holidaymakers would not recognize their idyllic sandy stretch on an open coast after a winter storm has scoured out all the sand. Fortunately, with time, it gradually builds up again before the next summer.

Where there are no cliffs backing a sandy beach, strong onshore winds will blow dry sand inland to form dunes. When a river finally reaches the coast and deposits its silt load, it accumulates to form muddy tidal flats riddled with winding creeks.

MORE OR LESS SALT

Estuaries are where sea, land and fresh water meet. The fresh water flowing out from the river mixes with the sea water to give a range of salt concentrations which vary at any one place with the ebb and flow of the tide.

◀During the high ranging spring tides, the incoming sea water builds up in the Severn estuary to form a wave-like bore.

▶The green alga *Enteromorpha* growing in a freshwater stream flowing down a shore.

▼Some of the salt pans at Salin-de-Giraud, which is the largest salt pan system in Europe, before they crystallize out and the salt is harvested at the end of the summer. The pink colour is due to organisms which tolerate living in this highly saline water.

The estuarine environment

Present-day estuaries have a variety of geological origins. Many are drowned river valleys, or rias, which have become flooded as either the sea level rose or the land subsided (p. 92). Sand bars or shingle spits deposited across a bay may dam up a river mouth to create a large brackish water area. The valleys gouged out by glaciation on the west coast of Scotland and Norway form fjords which are really gigantic estuaries. Since the last glaciation, the sea level has risen substantially so that the geographical positions of present-day estuaries date back only about 3,000 years.

Estuaries are places where the salinity, the water level, the amount of deposit and the force of the current all vary. These highly variable physical conditions greatly restrict the species of animals and plants that can dwell here. Water movements in estuaries depend on the height of the tide, the volume of freshwater outflow, the concentration of salts dissolved in the sea water and the wind force. The fastest currents are usually in the middle of an estuary where the water flow is least restricted by bank friction.

Estuarine water is often very turbid. Silts carried down by the river are kept in suspension by the tidal currents and deposited in the mouth of the estuary during the ebb tide (pp. 100–1). In suspension, these silts discolour the water and limit the penetration of light; hence limiting the photosynthetic activity of the plants.

Sea water flowing into an estuary is diluted by the river water to give brackish water containing much less than the 35 parts per thousand salt of normal sea water. The distance to which brackish water extends up an estuary depends on the channel size and profile and on the volume of fresh water. Salt and fresh water, however, do not always mix to form a gradual gradient. The fresh water tends to float above the denser sea water, so that the movement of fresh water out to the sea is counterbalanced by the movement of sea water up the bottom of the estuary

to form a salt wedge estuary. The rotation of the earth causes a tilting of the wedge, so in the Northern Hemisphere the fresh water flows out along the right-hand edge of an estuary, and the inflowing salt water hugs the opposite edge.

The typical pattern of semi-diurnal tides (pp. 100–1) is often modified in estuaries. For instance, in the Severn estuary (pp. 62–3), the funnelling effect not only causes an unusually large tidal range (14 metres at spring equinoctial tides), but also makes the rising spring tides form a spectacular tidal wave or bore.

When the numbers of species living in an estuary are plotted against the salinity, the number of marine species drops sharply as the mean salinity decreases, and the number of freshwater species declines at an even faster rate with increased salinities. Only a few species have adapted to a wide salinity range (often referred to as brackish water species) but since they face less competition than on open coast shores, they are often present in very

large numbers. However, the distribution of a species is by no means purely limited by salinity. Other factors can be of equal, or even greater, importance. It is now known that the particle size of the sediments is the prime limiting factor for *Corophium* (pp. 112–13); whereas the Baltic tellin is a deposit-feeding bivalve which is restricted to deposits with a high organic content.

Marine invertebrates which are successful colonizers of estuaries react to salinity changes by simply taking avoiding action; by migrating up and down with the tide (plankton); or by controlling their internal salt-water balance (osmo-regulation). Avoidance techniques include closing the shell valves (mussel); withdrawing into its shell (edible winkles); retreating into its burrow (ragworm); and withdrawing its siphons (peppery furrow shell).

The shore crab typical of rocky shores is also abundant in estuaries. It survives in salinities down to 6 parts per thousand by keeping its blood concentration higher (hyper-osmotic)

▲A salt 'mountain' at Salin-de-Giraud in the Camargue.

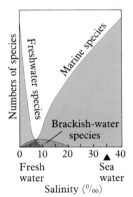

◄Graph showing how the relative numbers of freshwater, brackish water and marine species change with the salinity gradient of an estuary.

Graph labels: Numbers of species / Freshwater species / Marine species / Brackish-water species / 0 10 20 30 ▲ 40 / Fresh water / Sea water / Salinity (⁰/₀₀)

than the surrounding water, whereas the shrimp keeps its blood concentration lower (hypo-osmotic) than the surrounding water in salinities of 21–30 parts per thousand. The eggs and larvae of many marine species are much less tolerant of low salinities than the adults; for example, the shore crab eggs will not develop in salinities below 28 parts per thousand. Adult edible winkles will tolerate 7–8 parts per thousand, whereas the eggs will tolerate only 15 parts per thousand. To overcome this problem, several marine species living in estuaries migrate back to the sea to breed.

High salinities

In salt lakes and salt pans where the normal salinity of sea water is increased by evaporation, even fewer species are able to survive. One highly successful invertebrate is the brine shrimp, which is now extinct in Britain. This primitive crustacean, which is related to the freshwater fairy shrimp (pp. 36–7), will tolerate salinities from 3.5 parts per thousand to a fully saturated solution of

sodium chloride.

An extensive area of commercial salt pans exist in the south-eastern part of the Camargue in France. Sea water which is pumped in at one end of a line of shallow lagoons is gradually channelled from one to another, becoming progressively more saline. Finally, the saturated salt solution crystallizes out to a depth of 10 cm, when it is lifted off, washed and stacked into huge salt mountains. This salt-pan system at Salin-de-Giraud is the largest in Europe. The brine shrimps thrive in the salt pans, where they survive droughts as cysts. Flamingos and avocets can be seen feeding on live food in the intake lagoons, but they do not feed in the highly saline lagoons.

The Dead Sea is an inland lake where evaporation exceeds the freshwater input so that its salinity is approximately seven times that of the open sea. In recent years, however, even more saline water (320 parts per thousand) has been discovered in localized areas in deep rift valleys at the bottom of the Red Sea, where the water is also very hot.

Low salinities

In brackish water locations where the currents and variation in salinity do not show marked diurnal changes, a few marine and freshwater species are able to live side by side. In a shallow sea loch in north Scotland, pondweeds grow alongside marine mussels, and in the Baltic the water crowfoot *Ranunculus baudotii* intermingles with bladder wrack.

The route of a freshwater stream running across a flattish beach covered with rocks is usually clearly defined by a green ribbon of algal growth. When the streams run through gravel or sand with clean stones, the undersides of these stones are covered by the flatworm *Procerodes ulvae* and by the annelid worm *Protodrilus flavocapitatus*. Adapted to survive where they are alternately washed by fresh and salt water, they can escape from predation and suffer little competition from other animals.

The Baltic is a shallow sea which extends from Copenhagen to Finland and is so diluted by the 250 rivers which feed it that much of it has a salinity of less than eight parts per thousand. When sea water enters the Baltic it sinks, and the freshwater outflow moves out in the surface waters. Relatively few species live in the Baltic, and the distribution of those that do is closely linked to their salinity tolerances. The negligible tidal movement means that the salinity gradients are much more stable than in an estuary, so that some species can survive in even lower salinities; however, the annual freezing over of much of the Baltic creates an extra hazard.

PROJECTS

1 Buy some brine shrimp eggs from a pet shop (these are sometimes advertised as sea monkeys!). Make up a salt-water solution by adding a dessertspoonful of coarse cooking salt to about half a litre of water and stirring vigorously. Alternatively, if you live near the coast, you can collect your own sea water. Warm up the water (preferably with an aquarium heater) to 24–27°C. Do *not* put the container on a radiator. Sprinkle some eggs on to the sea water. After 24–36 hours the first stage nauplius larvae will hatch. If they are fed with Liquifry (available from pet shops) the larvae will moult several times and develop into adult brine shrimps. How long after hatching does it take for the adults to pair up? How does the male hold the female, and which way up do they swim? Compare the shape and size with the fairy shrimps on pp. 36–7.

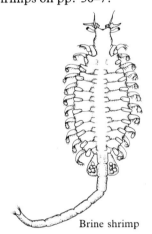

Brine shrimp

2 Collect a few freshwater shrimps and some water from an estuary. Place a shrimp upside down, held in a grooved piece of plasticine, and cover it with pure sea water (see above). Leave for 5 minutes and then count the number of times per minute that the amphipod beats its pleopods (A). Repeat several times, using different individuals, record your results and take the average. What do you think is the function of this beating? Does the number of beats per minute change with the salinity of the water? Do you think this could be the way in which these animals are able to adapt to varying salinities?

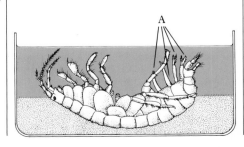

A

ESTUARIES

This peaceful, mid-summer scene belies the harsh, variable conditions of life in estuarine water, which contains a small range of species that are able to live there permanently. Taken during low tide, when all the estuarine water had drained out to the sea, this picture shows Gillan Creek in Cornwall, in the south-west of England, with the extensive mud flats and the main river channel on the left flowing down towards the sea. On the right of the exposed mud, a flock of herring gulls is feeding. The invertebrate animals which live in estuarine mud flats, especially in large bays, provide an important source of food for visiting waders during the winter months, huge flocks of which may be seen in some of the larger estuaries. The estuary is a drowned river valley, or ria, which is typical of this area of Britain. Rias were formed when river valleys were submerged as the sea level rose at the end of the last Ice Age. The regular tree-line, bordering each bank of the estuary, results from the salt-pruning effect of the brackish water.

A drowned river valley, or ria, in the south-west of England, showing the mud flats exposed at low tide and the trees extending down on both sides of the estuary.

STABILIZING THE MUD

Every day the incoming tide covers the salt marsh, damming up the river water into a lagoon where, in the still water of slack tide, the mud drops out of the water.

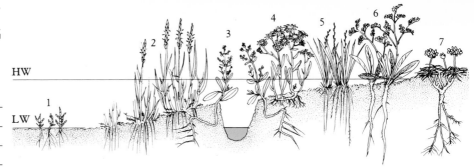

Section across a salt marsh from low water to high water mark, showing deposition of estuarine mud and erosion in the channels, with a succession of plants from those that can tolerate immersion in salt water to those that are only occasionally covered at the highest spring tides.

1 Glasswort *Salicornia europaea*
2 Cord grass *Spartina* x *townsendii*
3 Sea purslane *Halimione portulacoides*
4 Sea aster *Aster tripolium*
5 Sea arrow-grass *Triglochin maritima*
6 Sea lavender *Limonium vulgare*
7 Sea thrift *Armeria maritima*

Walking across the flat expanse of a salt marsh at low tide is like trying to walk through custard on jelly. You learn to keep to the areas that already have a cover of vegetation, where there is some stability in the ooze beneath you; but sooner or later you come across a deep channel, scoured in the mud by the tide, with steep sides of soft mud and a trickle of water in the bottom. On the sea side the salt marsh is protected from wave action by a shingle ridge or a sand dune system, and on the river side any scouring action is limited to the main river channel visible at low tide. So the mud accumulates, and the plants grow through it, giving us the characteristic salt-marsh landscape of vegetated flats fading into mud, with deep channels and salt-water pools left by the receding tide.

The rich organic mud, which has been collected from the whole catchment area of the river, provides abundant food resources for plant and animal life; yet it also presents problems to plants living in it. The mud covers everything indiscriminately, so only plants with a smooth surface that rejects the mud can continue to photosynthesize. When the water evaporates from the soil at low tide, it leaves a coating of salt, as well as pools where the water is highly saline. On a hot day this presents a problem to plants when evaporation from the leaf surfaces may exceed water uptake in a concentrated salt solution. However, at deeper levels in the mud the salt concentration does not vary much. At high tide the upper layers of the mud are very unstable as the tide regularly moves the muddy 'custard' around.

Plants that live in these conditions are called halophytes, meaning 'salt-loving plants'. If you taste a piece of glasswort (poor man's asparagus), it has a strong salty taste; the sap of many salt-marsh plants is more concentrated than land plants so they can absorb fresh water from their environment by osmosis (see pp. 10–11). To do this the concentration of their cell sap must be higher than the salt solution around

◄Salt pans on the salt marsh at Blakeney on the east coast of England contain water of high salinity due to evaporation. Glasswort *Salicornia europaea* grows on the margins and the perennial woody sea lavender *Limonium vulgare* on the stable mud between, where the salt content is not so high.

◄A muddy estuarine shore, showing the mosaic of conditions and vegetation patterns associated with erosion and the building up of hummocks in the lower salt marsh.

▼A sward of a cord grass *Spartina x townsendii* colonizing the lower zone of a salt marsh. The hybrid is fertile and will set seed after the pollination of these pale, brush-like stigmas.

Sea purslane *Halimione portulacoides*, a shrubby perennial that grows on the mud at the edges of drainage channels. The patches of bare mud in the foreground are colonized by glasswort *Salicornia europaea*, a fleshy annual that appears each year.

▼The red fescue zone of the upper salt marsh provides rich pasture for sheep in North Uist, Scotland. The muddy drainage channels extend throughout the salt marsh.

① Demonstrate osmotic pressure by tying a pig's bladder firmly over the wider end of an opaque plastic funnel. Fill it with concentrated sugar solution coloured with a little red ink, so that you can distinguish the level. Immerse it in a glass tumbler or jug of water as shown. What happens to the water level? Repeat the experiment but use a concentrated salt solution in the tumbler. What happens?

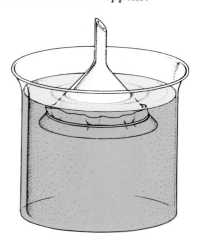

② Test the effect of different concentrations of salt solution on meadow and salt-marsh plants. Make a saturated solution of salt – that is, where no more water will dissolve in the tumbler. Make up different dilutions of it in labelled pots, using ranges of 1 part salt to 9 parts water, to 9 parts salt to 1 part water. Try placing a common meadow plant and a common salt-marsh plant in each pot, and take the time for wilting to occur in the different salt concentrations. Can you reverse the wilting process by placing the plants in fresh water?

③ Test the stability of mud in different areas of the marshes, using a graduated stick weighted at one end. Drop this from a fixed height and note how far it falls into the mud. What determines the softness of the mud?

them. Perennial salt-marsh plants will root at a deep level in the marsh where the fluctuations in salt concentration are not so great. However, glasswort is an annual with a much shallower rooting system, so it must be able to obtain water from a concentrated salt solution. Many of the salt-marsh plants have affinities with xerophytes (plants needing very little water) growing in desert conditions, such as reduced or fleshy leaves, or a tough, hard cuticle that resists evaporation or collapse, and an extensive rooting system. Establishment of seedlings in this unstable mud is a difficult process, and many glasswort seedlings are lost before their roots can anchor in the mud. Rice-grass is also a colonizer of the open mud, with rapidly growing underground stems, a fine network of feeding roots in the new mud layer, and many vertical anchoring roots. The hybrid rice-grass that first occurred in Southampton Water in the south of England in 1870 has now spread to many parts of the world, and is often planted as a mud binder in estuarine areas.

Once the mud is stabilized other plants move in, but the zonation that can be observed on the salt marsh relates to the tidal pattern and the length of time that each zone of

vegetation is covered by the sea. Late in summer the zones can readily be distinguished by the colours of their flowers: purple and yellow sea asters give way to the paler lacy sea lavender, which is succeeded in the upper marsh by pink carpets of sea thrift, often in flower at the end of May. Inundation by sea water presents problems of fertilization for flowering plants, so many salt-marsh plants flower late in the summer when the spring tides no longer cover the marsh and wind and insect pollination can occur.

At the edges of the creeks the grey-white plants of sea purslane, growing luxuriantly on a slight ridge above the level of the marsh, can often be found. The ridge is caused by the trapping of mud brought in with the incoming tide as it overspills from the muddy creeks. The sea purslane grows well on these ridges and rapidly excludes other plants in any zone, so that, seen from the air, these channels are fringed with pale grey.

Where the upper regions of a salt marsh are extensive, but rarely flooded, there may be a salt-marsh meadow of red fescue. This is traditionally valuable grazing, as the nutrient value of these old salt marshes is very high, owing to the organic content of the estuarine mud on which it is formed.

MOVING THROUGH ESTUARIES

Estuaries are habitats where physical conditions such as salinity, rate of water flow and degree of turbidity constantly change. Most animals cannot tolerate the complete change from fresh to sea water, but a few make remarkable migrations through estuaries.

The eel and the salmon are two fish which migrate back and forth through estuaries on their way to the feeding and breeding grounds. The salmon, which moves up rivers to spawn, is known as an 'anadromous' fish, while the eel, which moves down rivers to spawn in the sea, is a 'catadromous' fish.

Salt-water balance

Both these migratory fish therefore have to control their internal salt and water balance through a salinity range from less than 0.5 parts per thousand (fresh water) to approximately 35 parts per thousand (sea water). The concentration of the blood salts of bony fish is between that of sea water and that of fresh water, so, when fish are in the latter, they will tend to take up water and thereby swell. To counteract this intake of excess water, fish excrete a large volume of dilute urine when they are in fresh water. When fish enter the sea, however, the concentration of their blood salts is less than that of sea water, so they will tend to lose water, thereby becoming dehydrated. The fish then drink a great deal of sea water and also excrete chloride salts using secretory cells on their gills.

Besides the salmon, which is described in some detail below, the sea lamprey, the river lamprey, the sturgeon, the sea trout, shads and the smelt all migrate up rivers to breed.

Atlantic salmon

In summer and through to late autumn, a sharp rise in the water level of many of the clean, unpolluted rivers of Europe triggers off a salmon run. Fish which have gathered in the estuary start to swim rapidly upstream, leaping waterfalls up to 3 metres high. Not pausing to feed, except to snap at the super-stimulus of a fisherman's spinner

▲ Mature eels caught on their downward migration, on the grid of an eel trap.

► A salmon putcheon weir at Awre in the Severn estuary exposed at low tide, showing the conical putts.

▼ Salmon *Salmo salar* alevins four days old showing large yolk sacs which nourish the young salmon for the first few weeks of life.

▲ An elverer walks down to the bank of the Severn, with his net, early in the morning.

◄ Writhing glass eels or elvers of the European eel *Anguilla anguilla* are so transparent that their gills, heart and backbone can be clearly seen.

or fly, they gather in the headwaters where there are gravel reaches covered with about half a metre of water flowing at a rate of at least 10 cm per second. In November and December, the female, or hen fish, starts to cut a depression, or redd, in the gravel by flexing her body backwards and forwards, while the male, or cock fish, quivers alongside her. Once the redd is about 15–30 cm deep, the pair spawns. The hen fish then swims a little upstream and sweeps loose gravel with her tail over the eggs and starts the spawning sequence again. Each female lays up to a thousand eggs.

The exhausted, spent fish, or kelts, drift tail-first back to the sea, having lost 40 per cent of their body weight since they left it. Many die on the journey. Fish that do return to the sea quickly recover and start to feed again.

The eggs hatch in the spring as 'alevins', which are dependent on their yolk for about six weeks. They then start to feed on live food, and are known as 'fingerlings', or 'fry'. After a year, they develop eight to ten separate dark grey blotches down the side of the body and they are known as 'parr'. Trout and salmon parr look similar, but trout have more blotches, and have an orange, rather than a grey-green, adipose fin (the small fin on the back between the tail and the big dorsal fin). After two or more years, when the fish are 10–20 cm long, their flanks become silvery and they are called 'smolts'. In the spring, when the river temperatures begin to rise, the smolts move out into the estuary, where they pause for a while to get used to the salt water and a new diet of crustaceans and small fish. Then they

▲ Map showing the feeding area of the Atlantic salmon *Salmo salar*.

▶ Successive stages in the life history of the Atlantic salmon.

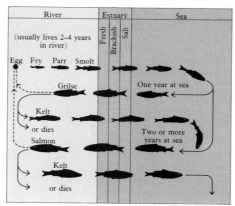

River				Estuary		Sea
(usually lives 2–4 years in river)				Fresh	Brackish / Salt	
Egg	Fry	Parr	Smolt			
		Grilse				One year at sea
	Kelt					
	or dies					Two or more years at sea
	Salmon					
	Kelt					
	or dies					

▲ Map showing the migration eastwards of the eel larvae from their spawning area in the Sargasso Sea.

▶ Successive growth stages in the life history of the European eel *Anguilla anguilla*.

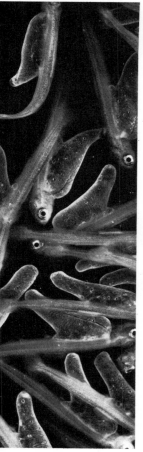

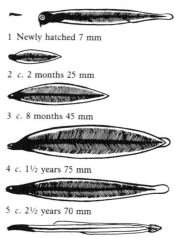

1 Newly hatched 7 mm

2 *c.* 2 months 25 mm

3 *c.* 8 months 45 mm

4 *c.* 1½ years 75 mm

5 *c.* 2½ years 70 mm

6 Elver *c.* 3 years 65 mm

the eel eggs. But it took Johannes Schmidt, a Danish marine biologist, fourteen years (from 1906 to 1920) to locate the position of the breeding ground of the European eel, by laboriously collecting and measuring Leptocephalus larvae from all over the north Atlantic Ocean. By plotting the size of the larvae from each sample, he built up a picture not unlike the isobars indicating a low pressure area on a weather map. The largest larvae were collected in the eastern part of the Atlantic and the smallest in a concentrated area in the Sargasso Sea to the east of North America. He concluded that this was the region where the adult eels came to breed; but final proof will come only when biological oceanographers see the eels spawning for themselves.

The Leptocephalus larvae spend some two and a half to three years being carried in the North Atlantic Drift current towards Europe. In the winter before they reach the coast, they change into transparent elvers, or 'glass eels', and the bodies become cylindrical. As the thyroid glands become more active, the elvers swim towards currents and hence up rivers.

Estuarine fisheries

Numerous local fisheries arose as a result of the regular annual migrations made by eel and salmon through estuaries, but many are no longer commercially viable. Seine netting for salmon is still quite widespread, but only in the Severn estuary in south-west England can the salmon lave-nets and putcheon weirs and the large elver nets be seen. There are now only a few lave-net fishermen still at work. The curious net, which is attached to a large, Y-shaped frame, is held in the hand, plunged into the water and shovelled forward before the salmon escapes. To be successful, these fishermen must have a quick eye and a rapid reaction as well as good instinct. Putcheons, or 'putts', are funnel-shaped wicker baskets which are erected as a weir so that the mouths of the baskets face downstream. Salmon trapped in the narrow end of the putts are collected by hand when the tide goes down and the weir is exposed to the air.

Elvering takes place in the Severn during the spring, when the young eels, or elvers, are caught in special boat-shaped elver nets, usually at night, when the elvers are most active. The net is held so that the mouth faces out to sea, and the elvers, which are swimming against the current, swim into the net. They are so slippery that they are simply poured out of the net into a container. Some elvers are eaten, but the majority are exported live for stocking European and Japanese inland waters.

undergo long seaward migrations, most of them journeying to the south-west coast of Greenland, where they grow very rapidly to 1½–3½ kg in the first year, 4–8 kg in the second, and 8–13 kg in the third.

After one to four years in the sea, the fish return to their rivers. By tagging salmon smolts on their downward journeys to the sea it has been proved that they are able to navigate back to the mouth of the same river by using an innate compass. They then find their way back to exactly the same headstream, using their sense of smell to detect the scent of its waters.

European eels

Eels spend many years (seven to twelve for males, nine to nineteen for females) feeding and growing in rivers. When mature, they move down towards the sea, especially on dark, moonless nights after heavy rain at the end of the summer. The normal yellow belly turns silvery and the sense organs – the nostrils, eyes and lateral lines – enlarge. At this stage, the gut begins to break down, so the eels are no longer able to feed.

Very rarely have eels in their breeding dress been caught at sea, and the location of their breeding ground is based on the collection of the flattened, leaf-like Leptocephalus larvae. Although these larvae were discovered in 1856, it was not until some forty years later, when two Italian scientists watched larvae in an aquarium change into young eels, that it was realized that they were not a separate species of fish but the stage which hatched out from

THE WAY OF A WADER

The North Sea coasts, particularly the estuaries, provide important stopovers on migration routes as well as overwintering sites for a very high proportion of the world's population of waders and wildfowl.

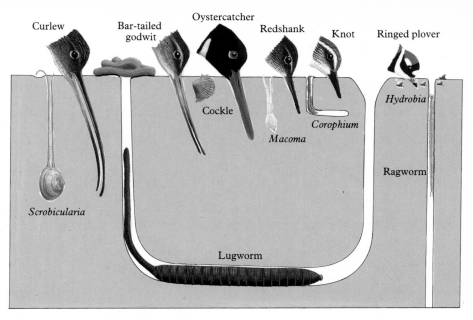

The general term 'waders' encompasses many species of bird which feed on saltings and mud flats, often in large flocks. They range from the large waders, such as the oystercatcher, the black-winged stilt and the avocet, to small species, such as the dunlin, plovers, sandpipers and stints. During the breeding season, waders radiate out into a wide range of habitats, which are often inland; in winter, they flock to estuaries and mud flats to feed on the rich concentrations of invertebrate food.

Getting around

To see a huge flock of small waders, such as knots or dunlins, change colour from white to brown as they reveal first their bellies and then their backs, by twisting and turning in mid-flight across an estuary, is an unforgettable experience, as is to hear the haunting calls of the waders.

When long-legged waders fly, they hold their legs outstretched behind their body as a counterbalance to the long bill. One of the most bizarre birds must surely be the flamingo, with a neck and legs which are proportionally longer than those of any other bird. As it flies, with slow wing flaps, for all the world like a multi-coloured flying coat hanger, the brilliant splashes of red, white and black are revealed.

Not only are waders able to fly fast, they can also run quickly and, if need be, swim. Short-legged plovers and sandpipers run rapidly for short periods and then 'freeze'. This behaviour, coupled with their cryptic coloration, makes it difficult to spot them against the muddy backcloth.

Food galore

In estuaries and saline pools, waders feed mainly on marine worms, crustaceans and molluscs; when inland, they feed on insects and earthworms. Waders do not eat plants, and they rarely eat fish. Short-legged waders have to wait for the ebbing tide to

expose the mud flats before they can feed, and they then spread out over the flats. High tide concentrates the birds into huge roosts on dry land, especially on isolated spits and points or in shallow water.

No two species of wader will feed on identical diets, or indeed, for the same length of time within a complete high/ low tide cycle; the dunlin spends about 75 per cent of the tidal cycle feeding, whereas the curlew spends only 47 per cent. Since the number of daylight hours in temperate latitudes is much reduced during the winter, waders do need to feed during a low tide at night, even when the sky is overcast. Waders tend to aggregate where food supplies are most abundant. The Baltic tellin can reach concentrations of almost 6,000 per square metre, while as many as

▲ Diagram showing how the length of a wader's bill limits the depth to which it can probe for food in the mud.

◄ A redshank *Tringa totanus* feeding on an exposed estuarine mud flat.

► Flamingos *Phoenicopterus ruber* feeding at dawn in the Camargue.

◄ A black-winged stilt *Himantopus himantopus* in flight, showing the characteristic flight poise of a long-legged wader.

► Oystercatchers *Haematopus ostralegus* congregate on a higher ridge of ground as an estuary floods at high tide.

63,000 per square metre of the tiny amphipod *Corophium volutator* (pp. 112–13) have been found in the Dovey estuary in Wales.

By the end of the winter, the populations of estuarine invertebrates have dropped considerably, but, since the individuals then reach their maximum weight, a wader gets a bigger food package and can cut down on the number of prey items it has to consume.

While long-legged waders are clearly able to wade out into deeper water to feed, their length of leg may have evolved so as to allow the birds to see above tall vegetation on their breeding sites, or to facilitate take-off. The graceful, long-legged avocets and stilts frequent shallow lakes, marshy deltas and saline lagoons. The avocet feeds on small crustaceans and other

Curlew, Bar-tailed godwit, Oystercatcher, Redshank, Knot, Ringed plover, Cockle, Macoma, Corophium, Hydrobia, Ragworm, Scrobicularia, Lugworm

▼ Black-tailed godwits *Limosa limosa* lifting-off, showing the distinctive black tail bar, long legs and long bill.

invertebrates by sweeping its upcurved bill from side to side, while the stilt feeds on similar food by using a rapid plucking action.

A flamingo feeds in a curious way, by bending down its long neck so that the bill is upturned. As the head is swept from side to side, water is drawn in and the food (small crustaceans and molluscs as well as algae and diatoms) is filtered off by layers of bristles inside the bill, in a similar way to that in which baleen whales filter off the plankton (pp. 168–9).

For waders which feed on prey living in the mud, it is the bill length which determines how far they can reach down into the mud to feed. Long-billed black- and bar-tailed godwits and curlews can take lugworms from deep down in their burrows; waders with medium-sized bills can reach crustaceans, worms and molluscs (tellins) in shallow burrows; short-billed waders are able to feed on life only in the surface mud.

The oystercatcher by no means confines itself to a diet of oysters; it feeds on a range of other bivalve molluscs including cockles, mussels and scallops, as well as gastropods such as whelks and limpets. An oystercatcher will consume 40 per cent of its own weight in a day. This is equal to over 200 cockles taken during one daylight tide, which means that these birds can have a marked effect on commercial cockle beds (pp. 166–7). During daylight, the bird appears to distinguish marks on the beach surface made by the cockles and so is able to home in on its prey. By night, the bird encounters cockles by walking forward with the tip of its bill probing just below the surface. Once a cockle is located, the oystercatcher levers it out of the mud with its strong, stout bill, which then makes a hole in the shell by means of repeated powerful blows. Although often using the same method as for cockles, oystercatchers can feed on mussels without holing the shell, by deftly attacking an internal muscle when the shells are ajar and then prizing the valves apart by opening its bill. From observations made in Britain during the severe winter of 1962–3, when oystercatchers were unable to feed, it is now known that the wear and tear on the bill is compensated for by continual growth, for when the bill is not used it will actually increase in length.

Turnstones live up to their name by using their stout bills to forage beneath stones, often turning them over in the process. They will also toss aside seaweed in the strandline in their search for sandhoppers. The purple sandpiper, which also forages in the strandline, is one of many waders which build up their food reserves in readiness for the long migration northwards to their breeding grounds in the high Arctic.

Threats to waders

Repeated disturbance to feeding waders — by low-flying aircraft or thoughtless horse riders — means that they not only spend less time feeding, but they also use up energy by repeatedly lifting-off. As estuaries are threatened by proposed reclamation schemes and also industrial development, it is important that as much information as possible is gained about the food preferences of waders as well as their movements to and from estuaries. This is being carried out by extensive research using ringing programmes involving both professional and amateur ornithologists working all over Europe and also in North Africa.

PROJECTS

① Visit an estuary in winter as the tide is falling. Take binoculars and a bird identification field guide. Record in a field notebook the date, the time of high water, the weather and the times of your observations. Observe with binoculars which species begin to feed first as the mud flats are exposed. Which birds wade out into the water to feed? What food are they feeding on? With practice, you will be able to recognize the birds on sight without reference to the field guide, and then more time can be spent on observing detailed behaviour by using a telescope mounted on a tripod. This may enable you to contribute data for estuarine bird surveys.

A snipe *Gallinago gallinago* wading out to feed.

② Examine empty cockle and mussel shells to see how many have been attacked by oystercatchers. Both will have a large, rough-edged hole, and mussel shells will have been attacked along their bottom edge where the shell is weakest. Use binoculars to observe the number of blows an oystercatcher has to make in order to penetrate a shell.

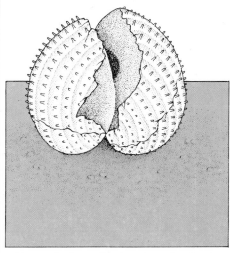

Damaged spiny cockle *Acanthocardia echinata* valves are proof of an oystercatcher attack.

LIFE IN THE MUD

From a distance, estuarine mud flats appear monotonously flat and lifeless, yet they support a rich invertebrate fauna which feeds casual year-round visitors as well as huge populations of winter wildfowl and waders.

Permanent inhabitants of estuarine muds have to tolerate fluctuating salinity levels; so, not surprisingly, the number of species which have adapted to survive in this harsh environment is limited. However, compared with sheltered open coast shores, they face much less competition for food and living space; so animals which are confined to estuaries may occur in very high densities in mud.

Living as a crowd

One of the most conspicuous animals exposed by the receding tide on estuarine mud flats is the snail *Hydrobia ulvae*, which is notable for its sheer abundance. A density of 60,000 per square metre (as found at Skalling, in Denmark) might seem to be a case of overcrowding, but the animals themselves are able to regulate their proximity to neighbours simply by moving further away. Each snail, which is a mere 4 mm long, was likened to a grain of wheat by Thomas Pennant, the eighteenth-century naturalist.

Immature individuals of the brackish water amphipod *Corophium volutator* will tolerate living in similar densities, although this is not apparent by walking over estuarine muds since *Corophium* soon crawls down into its burrow once it is exposed.

Surface clues

As on sandy beaches exposed by the receding tide, life in estuarine mud flats leaves clues on the surface as to its whereabouts below.

Young ragworms emerge from their burrows to feed on the surface mud. The lower half of the worm usually remains in the burrow, which aids a quick retreat if necessary. Irregular radiating tracks centred on the burrow hole are proof of the ragworm's presence. Less conspicuous irregular stars are made by the long feeding siphon of the peppery furrow shell as it sucks up detritus from the surface mud like a miniature vacuum cleaner.

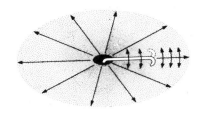

▲Diagram illustrating the feeding movements of the inhalant siphon of the peppery furrow shell *Scrobicularia plana*.

▶Proof of the hazard involved when man walks out over estuarine mud.

▲A young lugworm *Neanthes (Nereis) diversicolor* emerging from its burrow to feed.

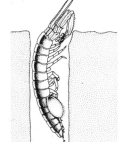

◀Diagram showing how *Corophium volutator* feeds by removing organic debris from the beach (A, B and C) and then retreats into its burrow.

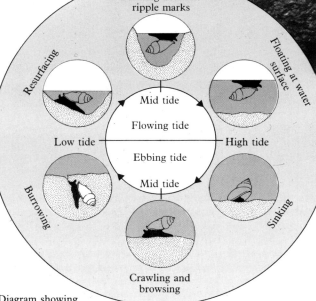

Floating between ripple marks

Floating at water surface

Resurfacing

Mid tide

Flowing tide

Low tide

High tide

Ebbing tide

Mid tide

Burrowing

Sinking

Crawling and browsing

▲Diagram showing how *Hydrobia ulvae* behaves on the shore during a complete tidal cycle.

thereby leaving broad scythe marks on the surface. As many as 3,000 snails have been found in the stomach of one bird.

Mud swallowers

When they are living in a sea of mud, it is not surprising that many of the inhabitants of estuarine mud flats feed on it, or rather on the organic detritus mixed with the fine mud particles. Indiscriminate mud swallowers either scoop, or suck, or engulf the mud. Baltic tellins and peppery furrow shells suck up the mud with their inhalant siphons.

Hydrobia snails have a distinct behaviour pattern which is related to the tidal cycle. When first exposed to the air, *Hydrobia* crawls over the mud engulfing the detritus and diatoms with its radula, then it burrows into the mud just below the surface. Before being submerged by the incoming tide, it resurfaces and crawls up a ripple mark, so that it can flip over and hang upside down on the surface film by a mucus raft. This raft keeps it afloat, and also traps plankton. While clinging to the raft, *Hydrobia* is buoyed up the shore by the incoming tide and eventually gets stranded in the mud again. This curious behaviour cycle, which is triggered by responses to light and to gravity, allows *Hydrobia* to feed throughout the tidal cycle.

Since young *Corophium* do not leave their burrows, they are probably able to filter off fine food particles from their incoming respiratory current. The more usual feeding method, however, is to select deposits from the surface. By lying upright in its burrow, *Corophium* can scrape organic debris off the surface with its second antennae. Long hairs, or setae, on the feeding limbs are then used to filter off the food.

Hunters

Detritus feeders fall prey to permanent estuarine inhabitants as well as to the hugh transitory bird populations. Ragworms will scavenge on detritus or dead animals as well as prey upon small crustaceans. The ubiquitous shore crab which is abundant on rocky shores also lives in estuaries, where it can tolerate salinity as low as 6 parts per thousand. While shore crabs will scavenge, they are also active predators and will feed on anything they manage to grab with their pincers.

The Chinese mitten crab, a native of east Asia, is so-named because of its furry pincers. This crab has been introduced to Europe, where the adults live in fresh water and migrate down estuaries to breed. They cause damage to estuarine banks by their burrowing activities and they will also damage fishing nets by cutting through the mesh.

Lugworms will also live in estuaries, where they make their familiar surface casts (pp. 122–3). Circular, or even figure-of-eight tracks in the muddy sand mark the trail of a winkle which maintains its tidal position on the shore by orientating itself in relation to the sun.

Inhabitants of the mud are not the only animals which leave such clues. Predatory birds which come to feed on them leave both walking and feeding tracks. During the winter in particular, wader footprints will be plastered all over the mud surface, only to be obliterated by the next incoming tide. As a snipe feeds on invertebrates, its long bill leaves an irregular line of deep probing holes. Shelduck use quite a different technique to collect *Hydrobia* snails: as they walk over the mud, they sweep their bills from side to side,

PROJECTS

1 Visit an estuary at low tide to look for surface clues left by mud inhabitants and their predators.

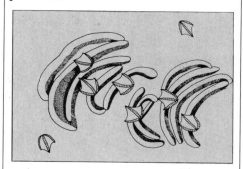

The scythe-like feeding marks made by a shelduck feeding on the surface mud.

2 Tie a piece of strong string to the handle of a large plastic bottle. Use this to take samples of estuarine water from a jetty or a high wall during the ebb and the flow stages of the tidal cycle. Shake up each sample before transferring it to a glass jar. Allow the sediments to settle and measure the depth of sediment in each sample.

3 Use cardboard strips 3 cm deep to divide a washing-up bowl into four equal-sized compartments. Add estuarine mud in two diagonally opposite compartments (A-A) and sand in the others (B-B) to a depth of 2 cm. Carefully remove the cardboard. Place 20 *Corophium* on the surface at the central intersection of the sediments. Put the bowl in a place where the temperature and lighting are constant. After 5 hours replace the cardboard dividers before carefully sorting through each compartment to discover how many *Corophium* are in each one. Is there any proof that the *Corophium* are selecting the grain size of one sediment type in preference to the other?

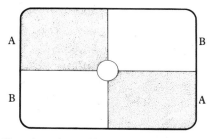

Plan view of a washing-up bowl divided into four compartments. *Corophium* are released on to the central area of the intersection.

Do not walk out over large muddy expanses without spreading the weight of your body by wearing snow shoes or boards attached to your boots.

NATURAL EXPLOITERS

Estuaries are vital winter refuges for huge numbers of migrant waders and wildfowl, as well as year-round havens for a few resident birds and fish.

The high density of invertebrates living in estuaries provides a rich source of food for many resident and migrant birds. As well as waders (pp. 110–11), herons, cormorants, various ducks and geese, gulls and terns all visit estuaries to feed. Some live there for most of the year, while others are temporary opportunist visitors.

Permanent residents

Mute swans are by no means confined to estuaries, but some individuals frequent the upper reaches and creeks where they nest on raised ground above high-water level. They move down to the water to feed on green algae and crabs. These heavy birds are well adapted for walking over fine muddy flats, with their broad, webbed feet spreading their weight.

Shelduck spend most of the year in estuaries, nesting on stabilized sand dunes or grazed marshes, often utilizing an old rabbit burrow. Breeding shelduck exhibit a strong territorial behaviour which disappears after the young ducklings hatch and are gathered together into crèches. A few adults remain as guardians, while many adults migrate in July to the Heligoland Bight, off north-west Germany, for the annual moult. All the flight feathers are then lost and so the shelduck cannot fly; they merely float up and down with the tide. Shelduck are able to feed during all stages of the tide either by wading out over the mud to feed, chiefly on *Hydrobia* (pp. 112–13), or by up-ending themselves in water. Both the ducks and the drakes have a striking black, white and chestnut plumage.

Flounders often spend most of their lives in an estuary, moving down to the sea only to breed. In February to May, a single fish lays 400,000 to 2,000,000 1mm-diameter eggs. After five to seven days the eggs hatch into transparent, 3mm-long larvae, which live as part of the plankton until mid-summer, when they sink to the bottom and metamorphose into tiny fish (pp. 128–9). The flounder is an important predator in the estuarine food pyramid,

▲ Mute swans *Cygnus olor* feeding on crabs in an estuary.

◄ A flounder *Platichthys flesus* undulates its body as well as its marginal fins to propel itself forward.

▼ Shelduck *Tadorna tadorna* feeding in an estuary in winter.

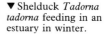

▼ A grey heron *Ardea cinerea* with a fish caught in the shallows on the rising tide.

▲ A juvenile glaucous-winged gull *Larus glaucescens* scavenging on an old fish-head beside fish docks on the Californian coast.

sharing prey such as ragworms, *Hydrobia*, Baltic tellins and *Corophium* with sand gobies, shelduck and several waders. Flounders are able to feed only when the tide submerges the mud flats, which means that they need to forage during both the day-time and the night-time high tides. They are caught on a commercial scale in traps, fixed nets and seine nets. The annual European catch is about 10,000 tonnes, and the fish are marketed both fresh and smoked.

Temporary visitors
Opportunist feeders, such as gulls, will move into estuaries to feed as predators or scavengers on any food which is available. They can be seen trampling up and down over a small area of the exposed mud to bring invertebrates up to the surface; and herring gulls will fly up with an intact bivalve in their bill so that it can be dropped from a height on to hard ground and the shell cracked open. Cormorants and herons seek out fish in estuarine pools and creeks.

Shovelers, mallard, wigeon, teal and pintails are all dabbling ducks which feed in shallow water or on the mud. Flocks of diving ducks such as eider, scaup, goldeneye and long-tailed ducks move into estuaries to feed when the tide turns.

Eider ducks breed around the shores of northern Europe on flat, rocky and sandy shores, where it is a commonplace sight to see several ducks with their fluffy ducklings swimming in sheltered lochs, voes and fjords. The female duck plucks out her breast down for lining the nest. In parts of the Arctic and sub-Arctic, nesting colonies of eiders are farmed so that the down can be collected for stuffing quilts. Large rafts of eider gather inshore in spring, prior to breeding, and also during the winter in large Scottish firths, including the Firth of Forth.

The thick-lipped grey mullet often frequents harbour mouths and estuaries, where shoals can be seen swimming just below the surface. It feeds by gulping in mouthfuls of mud and finer algae, passing out the mud through the gills and swallowing the algae, plus small worms and crustaceans. Along harbour walls and piers, grey mullet are caught on lines by fishermen.

Winter migrants
Estuaries all over Europe are important havens for huge concentrations of overwintering birds. Flocks of waders and large numbers of wildfowl – especially geese – migrate to estuaries, where they spend the winter months before moving northwards to breed.

All the Spitsbergen population of barnacle geese overwinter in the Solway Firth on the west coast of Scotland. They do not feed on the intertidal mud flats but on the salt marsh, or 'merse', as it is known locally. In October, all the pink-footed geese which breed in Iceland and east Greenland and all the Icelandic greylag geese arrive in Britain, where most of them overwinter in Scotland.

Brent geese are the only geese which survive the winter by feeding intertidally – preferably on eel grass or, failing that, on the green alga *Enteromorpha*. Early in the 1930s, British eel grass beds were wiped out by a mysterious disease, which brought about a drastic decline in the numbers of brent geese. Now that the eel grass beds have re-established themselves and it is illegal to shoot brent geese in Britain, the overwintering population has increased threefold, from under 30,000 in the late 1960s to 90,000 in 1977–8.

Estuarine surveys
The data obtained as a result of the Birds of Estuaries Enquiry in Britain (1970–5) and the Wetlands Enquiry in Ireland, have provided a great deal of information about the most popular estuaries as regards numbers of waders and wildfowl. These figures are vital for conservationists in their fight against the adoption of projects which would greatly change the tidal regime of an estuary (pp. 116–17), and hence the food available for these birds.

(pp. 116–17)

1 Use a European bird field guide to find out where whooper and Bewick's swans, pink-footed and white-fronted geese overwinter, and where they breed in summer. Mark the locations on a map. Estimate how far they travel between sites in a year.

Whooper swan *Cygnus cygnus* taking off along the water surface.

2 Find out where there is an estuarine reserve with a public hide. Visit it with a bird identification book and a pair of binoculars. Try to identify as many species as possible. Can you spot any behavioural patterns which are being repeated? Are any of the birds ringed? Visit an industrialized estuary. Are there more or fewer birds here than in an undisturbed estuary? Also, compare the number of species in the two sites.

The barnacle goose *Branta leucopsis* is one of several species of goose which can be seen overwintering in European estuaries.

3 Look at rocky patches of ground close to estuaries to see if you can find any evidence of gulls having dropped live bivalves. If you find an area with a lot of empty shells, crouch quietly nearby and observe, with a pair of binoculars, how the gulls feed. Time how long it takes for a gull to break a shell open.

CHANGING THE WATER

Now, more than at any other time, estuaries are at risk, threatened by development of industry, oil refineries, tanker terminals and barrages, as well as by land reclamation.

Estuaries are naturally very fertile habitats, which are much more productive over a given area than a wheat-field. This productivity is founded on the high activity of nitrogen-fixing bacteria in the muds enriching the estuary with nutrients. However, little of this productivity is carried out of the estuary; neither is it exploited to any extent by man. Unlike crops on land, which are grown and harvested on the same ground, marine foods may develop in one region and be harvested in another.

Poisonous heavy metals such as copper, cadmium, mercury, lead and zinc, which are discharged into estuaries from mining and chemical industries, can be accumulated by some invertebrates. It is now known that several estuarine molluscs are able to enclose heavy metals in vesicles, thereby isolating them, but a predator feeding on these molluscs may have no such protection.

Several of the larger estuaries are already the sites of power stations, which draw in water for cooling purposes and return it up to 12°C above the ambient estuarine water (pp. 170–1). Now, serious proposals are being put forward for the construction of estuarine barrages as a means of harnessing tidal power. The latter could have far-reaching long- and short-term effects on wildlife, and especially on the large populations of overwintering waders and wildfowl.

The Wadden Sea
The shallow waters of the Wadden Sea comprise the largest estuarine area in Europe. The region bordering the Dutch coast is an important nursery ground for the common shrimp, plaice, sole and herring, all of which are fished commercially in the North Sea. Millions of seabirds also visit the area to feed.

The western part of the Wadden Sea has relatively little pollution. However, effluents are discharged from industrial German cities, such as Bremerhaven and Hamburg, and from dairies and slaughterhouses in Denmark. All these discharges result in a marked local

oxygen deficiency. Fortunately, these effluents are soon diluted within the huge Wadden Sea, which is still the most productive estuarine ecosystem in Europe.

Power station spin-off
Warm water effluents from power stations are generally detrimental to life in estuaries, since they both reduce the oxygen-carrying capacity of the water and may subject life in the vicinity of the outflow to thermal shock.

Warm water effluents can, however, be used to accelerate the growth of commercial fish and shellfish which will tolerate the confines of enclosed ponds.

At the marine farm at Hinkley Point Nuclear Power Station on the Severn estuary, in the south-west of England, experiments were carried out over a period of several years to see which species kept best. Six species of prawn, including the large tiger prawn, and turbot, sole and Pacific oysters all proved easy to keep, but their production was not economic. Now, only eels are fattened up in the marine farm. They are fed on a readily available food supply: the fish caught on the screens of the cooling water intake to the power station. The fish, which are deep-frozen, are fed to the eels by mincing them up to form a paste.

◀Lifting out a tray of Pacific oysters (*Crassostrea gigas*) from a warm water pond at the Marine Farm, Hinkley Point Power Station.

▶A picture westwards of Kingsnorth Power Station on the river Medway, where the outfall water temperature may reach 10°C more than the intake water. This area is an important study site for warm water effluent research.

◀An infra-red line scan of a Rotterdam dock area, showing the extent of industrial discharge and hot water (white) into the main stream.

▶Part of the Central Electricity Generating Board's nature reserve at Connah's Quay on the Dee estuary, showing the salt-marsh habitat. Behind are the Environmental Centre, the electricity sub-station and the large cooling towers of the power station.

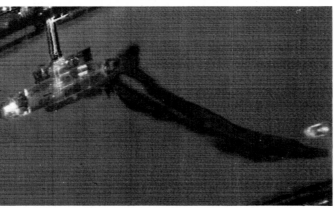

◀An infra-red scan showing oil pollution from a tanker moored in the Botlek dock area of Rotterdam.

Power station intake

Outfall lagoon

Direction of flow

Salt marsh

Salt marsh

Further north, at Connah's Quay Power Station on the Dee estuary, a 36-hectare nature reserve of open water, salt marsh and tidal mud flats has been developed along the shoreline, with the full co-operation of the Central Electricity Generating Board. The salt marsh area was reclaimed, using pulverized fuel ash, a waste product produced by burning coal to generate electricity. More recent management has included the construction of a marsh scrape to provide a high tide roost for wading birds. Even the electricity sub-station itself is attracting wildlife, for a pair of kestrels chose one of the struts as their nesting site!

The Severn Barrage Scheme

At present, the south-west of Britain is not well supplied by electric power from local power stations. The proposed Severn Barrage Scheme, in an estuary with the second highest tidal range in the world, could supply some 6 per cent of the present United Kingdom requirements for electricity, and some 50 per cent of the requirements for Devon and Cornwall. It would be possible to enclose the estuary by a barrage sited anywhere east of Porlock in Somerset, but the most likely siting for it is from Weston-super-Mare across to Cardiff on the Welsh coast.

What major changes would such a barrage bring about? It would alter the tidal regimes inland of it, reduce the huge tidal range from 12 metres to 3–6 metres at spring tides, and eliminate the Severn bore (pp. 102–3). Pollutants discharged into the estuary would be dispersed more slowly and silt would accumulate in channels, thereby restricting the passage of ships. The migrations of both eel and salmon (pp. 108–9) would obviously be affected. During the winter months large numbers of waders and shelduck visit the Severn estuary to feed on the intertidal mud flats, the extent of which would be much reduced behind a barrage. Wild geese also come to feed on the saltings at the Wildfowl Trust headquarters at Slimbridge in this area. As with any extensive man-made construction, however, the full effects of a barrage on the permanent and visiting wildlife of the Severn estuary will not be known until after it has been completed.

Salt pan encroachment

Ever since defensive dykes were built in the Camargue in the Middle Ages to prevent the Rhône from flooding farmland on the delta, man has made his mark on this wild European landscape. In more recent years, roads have been built and artificial dykes constructed, so that over 55 per cent of the area is now given up to agriculture and the rearing of cattle and wild horses. There has also been a dramatic encroachment of salt pans in the Rhône delta since the last century, which has greatly reduced the area of natural brackish water lagoons for wildlife. The Camargue is not only a picturesque wilderness area, but also an internationally important stopover place for migrating birds, an overwintering site for large numbers of duck in the brackish lagoons, and a breeding site for flamingos (pp. 110–11). Since salt pans support a very limited number of species, their extension greatly decreases the diversity of plants and animals.

The declaration of the 13,760-hectare Parc Régional de la Camargue in 1970, as well as several other reserves, will ensure the protection of this land from agricultural or salt-pan development, which has doubled in area within two decades in this century. This is not enough, however, to save the range of wildlife in this unique region. At the moment, private land adjoining the reserves used for grazing and hunting not only extends the wilderness areas but also increases the diversity of the habitats, not all of which are encompassed within the reserves. However, will the increasing world demand for salt result in the disappearance of these wild areas?

A CHANGING COAST

Major changes in the profile of the coastline result from rising and falling sea levels; while minor changes occur as a result of the erosion of cliffs and the deposition of sand or shingle elsewhere.

Fluctuating sea levels

As well as the daily variations in high water level resulting from the alternating cycles of spring and neap tides (pp. 100–1), much greater fluctuations in sea levels have taken place over prolonged periods. During the Ice Ages, when large quantities of water were frozen, the sea level was consequently lower and the weight of ice compressed the land. Icebergs, which broke off from the snouts of glaciers and floated out to sea, were grounded in what was then shallow water, and left deep scour marks that can still be seen at the edge of the continental shelves at latitudes down as far as south-west Ireland. When the ice began to thaw, some 10,000 years ago, the sea level took about 2,500 years to rise to the present level. As the sea rose, it submerged low-lying land, drowning river valleys to form rias (pp. 104–5). It also flooded the English Channel and the Dogger Bank, between England and Denmark, where trawlers sometimes haul up mammoth bones, proving that these areas were once dry land. It submerged coastal forests, remnants of which now exist as offshore submerged forests. Some of these can be seen today when they are exposed to the air at low tide. Gradual changes in sea levels (both rising and falling), in relation to the land, are said to be 'eustatic'.

Raised inland beaches, as much as 30 metres above the present sea level, are proof that the sea reached higher levels in the past. Such beaches are not uncommon on the north-west coast of Scotland. They became exposed as the ice melted and the land rose with the lifting weight. Uplifting of the land which takes place in this way is known as an 'isostatic' adjustment.

Sunken cities

Evidence of rising sea levels exists as submerged ruins on the sea bed. Early Stone Age caves and remains of Neolithic villages became submerged on what is now the European Continental Shelf when the sea level

rose at the end of the last Ice Age. Divers have found Stone Age sites off the coast of Gibraltar and Italy as well as more than a hundred sunken cities in the Mediterranean at depths of 1–5 metres. The modern coastline of Italy, from Naples to the west of Posilipo, is quite different from that of Roman times, when the sea was 5 metres lower and luxurious villas bordered a great highway at the base of the cliffs. The disappearance of these cities into relatively shallow depths has been largely due to gradual earth movements.

A drowned desert

Evidence gathered during the last twenty years, including work carried out by the deep-sea drilling vessel *Glomar Challenger*, suggests that six million years ago the Mediterranean basin was a huge desert lying some 3,000 metres below sea level. A vertical crack in the lower part of a deep-sea core sample, taken through 335 metres

of muds overlying an area west of Sardinia, is thought to be a desiccation crack, which is characteristic of drying-out sediments. Seismic profiles have revealed pillar-like structures (thought to be salt domes) beneath the western Mediterranean floor.

The dried-up Mediterranean would have been like a giant bath-tub, which became filled with Atlantic sea water breaking through the Strait of Gibraltar as a huge waterfall some five and a half million years ago. Calculations suggest that the rate of inflow must have been a hundred times that of the rate of flow over Victoria Falls and a thousand times greater than the Niagara Falls, otherwise the water lost by evaporation would have been greater than the infilling. Even so, it would still have taken more than a century to fill the empty bath-tub.

Eroding coasts

The powerful movement of the sea and the action of frost, rain and sun are

▲An aerial view of the curving spit of Spurn Head, off the north-east coast of England, shows the man-made groynes on the east of the spit, and the vegetated sand dunes.

◄A view along a raised shingle beach at Harris, on the Isle of Rum, off the west coast of Scotland.

►Tree trunks from a forest which was submerged by the sea are now exposed on a shore at low tide. They provide a site for green seaweeds to grow up above the beach.

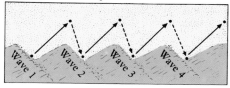

▼A diagram showing how longshore drift moves a pebble along the beach by the action of successive waves.

Top of beach

Wave 1 Wave 2 Wave 3 Wave 4

▲Man-made breakwaters built out at right angles to a sandy shore help to reduce the force of the waves undercutting the sand dunes, and also the longshore drift of the shingle.

constantly attacking the coastline. Soft cliffs formed of glacial deposits are eroded more rapidly than hard rocks (pp. 152–3). In the north of England, along the 56km stretch of the Yorkshire coastline from Flamborough Head to Spurn Head, at least twenty-five villages or hamlets have been lost to the sea since Roman times. The erosion of cliffs may give rise to interesting, and maybe dramatic, coastal features such as caves, arches and stacks.

Fossilized plants and animals, which have been laid down in sedimentary rocks along the coast, may be revealed as the sea continues to attack a wave-cut platform (pp. 100–1) or, more especially, when cliffs crumble away. Fluctuating sea levels can result in fossils of marine animals now occurring some distance inland.

Powerful wave action, especially during winter storms, can result in the removal of sand from coastal sites. It was Danish fishermen who noticed the calming effect that seaweed has on the sea, and thereby hit on the idea of using plastic seaweed to combat coastal erosion: 2.5-metre lengths of polypropylene rope, resembling thongweed, are attached at 1-metre intervals to lengths of steel chain laid in rows at right angles to the shore. Preliminary tests in sheltered Danish waters resulted in a build-up of sand to a depth of 1 metre over an eighteen-month period. This application of plastic seaweed could not only help to stabilize holiday resorts but also prevent the erosion and undermining of deep-water pipelines.

New coastlines

As coasts are being eroded in one place, the sand and shingle formed by the action of sea on the fragmented rock are deposited elsewhere to form bars, spits and tombolos.

Spurn Point in Yorkshire is a good example of the cyclic development of a sandy spit which has been well documented by historical evidence. The full cycle of formation and erosion of the spit takes about 250 years to complete. Erosion of the Holderness coast to the north is very closely linked with the development and the erosion of successive spits. Each new spit grows to the north-west of the previous one as the coast to the north is eroded away. As the neck of the spit is eroded away it is breached, so the end is carried across the Humber estuary to build up the north Lincolnshire marshes. Since the middle of the nineteenth century, the neck has been artificially strengthened by groynes, and the spit has continued to grow southwards with a wedge-shaped tip.

Harlech Castle was built on the top of cliffs on the mid-Wales coast with its own harbour. The development of a large, sandy spit has now caused the castle to be separated from the sea by nearly a kilometre. The great biblical port of Ephesus, in Turkey, is now several kilometres inland. This is the result of complete erosion of the topsoil, following the felling of forest cover on the hills bordering the Mediterranean, which deposited a vast quantity of sediments down into the sea.

As waves reach a long, open beach, they can usually be seen moving along the shore at an oblique angle to it. This means that shingle carried by the waves is not dumped in a neat, straight line at the top of the shore; instead it is moved diagonally up the beach, and, if there is a strong backwash (pp. 100–1), back down the beach at a right angle. The net result is that unless groynes intercept the natural lateral movement of the shingle it will be carried from one end of the beach to the other. Shingle piled high upon one side of a series of groynes or breakwaters is a clear demonstration that this longshore drift is taking place.

PROJECTS

1 When at the coast, visit a museum and look for a local fossil collection. This may provide clues to any beaches likely to be productive for fossil hunting. Fossils occur in most *sedimentary* rocks, formed by deposits building up on sea or lake beds. You will be able to find some fossils simply by examining rocks exposed during low tide. In other sites, it may be necessary to use a geological hammer and a chisel to split rocks, but this should be done only with the guidance of a person who has some knowledge about fossils.
Never try to break up or to climb crumbling cliffs to look for fossils.

The fossilized internal skeletons of belemnites, the forerunners of squids, exposed at low tide on a wave-cut platform on the Algarve coast in Portugal.

2 Collect 6 even-sized pebbles from a shingle beach. Wash them in fresh water, dry them, and paint them all over with white emulsion paint. After the paint has dried, drop the pebbles into the surf and time the speed at which they move up and along the beach.

SAND AND SHINGLE

The large size of the boulders on this offshore island in New Zealand, rounded by being continually rolled by the sea, testifies to the violence of the wave action which can occur on this exposed beach. Any visitors to the island must land here and more often than not the rough seas prohibit a safe landing. In such a violent habitat relatively few types of animals or plants can survive, as they continually run the danger of being ground between the rolling boulders. Those that can survive include encrusting lichens on the uppermost boulders. On this boulder beach any wood-boring animal able to exploit wood will never be short of food. Beaches that are less dynamic contain progressively more variety of animals as the physical environment becomes less rigorous. The size of the component particles of a sand or shingle beach gives a very good indication of the wave action needed to move the beach material up on to the shore: the larger the sand grains or pebbles, the more exposed the beach to heavy swell. A boulder beach is at the extreme end of an exposure scale of a shore formed by the deposition of particles eroded from land elsewhere; whereas fine, silty mud can accumulate only in sheltered inlets free from wave action.

A boulder beach on Little Barrier Island, New Zealand.

BURROWING DOWN

Unlike rocky shores, sandy beaches do not support such a wide range of microhabitats, and so fewer species tend to occur. However, those which are well adapted for living and feeding in sand can be very numerous.

In the height of summer, the most obvious animals on most beaches will be the holiday-makers; but these temporary visitors are far outnumbered by the permanent inhabitants in the beach itself. The way in which sand originates is not significant to sandy beach fauna, whereas the ultimate grain size is crucial. The sorting of particles by the swash and backwash of waves means that the larger particles tend to get carried further up the beach, while the finer particles are dragged back to the lower beach. Only a few centimetres down, beneath this finer sand, which retains water well when the beach is exposed, the temperature and salinity remain fairly constant. Here, at low tide, the sand infauna sits and waits until it can start to feed again as soon as the tide floods the shore.

Why burrow?

If the inhabitants of the intertidal zone of sandy beaches are to survive during low tide, they must burrow into the sand before the beach is exposed by the ebbing tide; otherwise they would be susceptible to damage or death by wave action, desiccation and temperature extremes, as well as the risk of predation. On flat sandy beaches there are no rocks or seaweeds nearby for cover, so any animal which is left stranded on the sand is easy prey for sharp-eyed gulls or oystercatchers.

As well as being able to escape by moving quickly downwards, burrowing animals must also be able to extricate themselves if they should be covered with too much sand by wave action. Worms can do this simply by turning round, while bivalve molluscs push downwards with their feet so that their shells move upwards. In this way, some cockles can make spectacular leaps right out of the sand.

Dark green patches are not uncommon on coarse sandy beaches on the north Brittany (in north-west France) and Channel Island coasts.

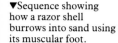

▲A female masked crab *Corystes cassivelaunus* beginning to dig its way down into the sand.

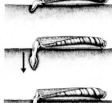

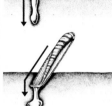

▼Sequence showing how a razor shell burrows into sand using its muscular foot.

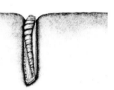

▶A spiny cockle *Acanthocardia echinata* with its large pink foot extended from between the valves.

These are large colonies of the marine flatworm *Convoluta roscoffensis* which move up on to the beach surface as soon as it is exposed to the air so that their symbiotic algae (pp. 148–9) can photosynthesize. The worms react to the vibrations of the waves moving back up the shore by burrowing down again and thereby preventing themselves from being washed away.

Methods of digging

Animals which burrow may have a hard body, such as the masked crab; a completely soft, worm-like body; or, in the case of bivalve molluscs, a combination of a hard exterior and a soft, burrowing foot.

Hard-bodied animals burrow using several appendages, for example, the five arms of the burrowing brittle star

and all four pairs of walking legs of the masked crab. Occasionally, a masked crab can be found exposed low down on the beach. If you fill a bucket or jar two-thirds full of sand, top it up with sea water and put in the crab, the rapid way in which it burrows can be seen. Sitting up vertically, it digs down by forward and upward movements of the legs until only the tips of the antennae are left projecting above the sand. The two rows of hairs on the inside faces of the antennae interlock to form a breathing tube enabling the crab to draw down a current of fresh sea water.

Starfish and sea urchins are more typical inhabitants of rocky shores; but the burrowing starfish and the burrowing heart urchin are both well adapted for their life in sand. Instead of having tube feet ending in suckers, the

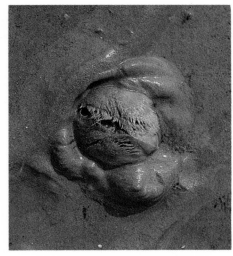

▲A lugworm *Arenicola marina* starting to burrow into sand.

◄Marine flatworms *Convoluta roscoffensis* are coloured green by symbiotic algae living inside them. The flatworms undergo a simple but regular tidal rhythm by moving up on to the sand when the beach is exposed and burrowing down again before the incoming tide reaches them.

underside of the test, the specialized, spatula-like spines are used for digging.

Many of the soft-bodied, burrowing animals are circular in cross-section, including polychaete worms, the burrowing sea cucumber and the burrowing sea anemone. Their body cavity functions as a hydraulic organ, by pumping liquid from one part of the body to another. Worm-like animals burrow by a series of alternating waves of contraction and expansion of the burrowing starfish has pointed tube feet which push away the sand grains so that the animal gradually sinks downwards. Like all sea urchins, the heart urchin has a rigid outer test (illustrated on p. 132); but, instead of having hard, pointed spines, it has soft, golden spines which face the same way so that they lie back on the test. On the

body: the front end is first forced down into the sand and then rapidly dilated so that it prevents the animal from moving backwards and upwards. A terminal anchor is then formed by the body wall pressing against the sand so that the body can be pulled downwards.

In finer sand, particularly if it also contains some mud, lugworms abound. These worms are taken by sea anglers for bait; they are also preyed upon by seabirds. The tugging action of a predatory gull on the tail of a burrowing lugworm causes the front end of the worm to expand so that it grips its burrow even more tightly. Additional anchorage is provided by bristles which project from each segment of the body. If necessary, a lugworm can escape by shedding its tail, which is then renewed.

Bivalve molluscs have to work harder at digging, because they have to drag their shells down with them. This is especially true in sandy beaches, where the bivalves tend to have slim profiles (e.g. the banded wedge shell) compared with mud-dwelling bivalves such as the sand gaper.

Even though the shells of a razor may reach up to 20 cm long, they are very light and their smooth sides offer little resistance to burrowing. Only after a storm are live razor shells likely to be exposed on the shore, for they normally remain buried in sand. If a razor shell is dug up, however, it will immediately start to burrow by inserting its wedge-shaped foot into the sand. As blood is pumped into the end, it swells and acts as an anchor, so that its two shells can be pulled down. This sequence is then repeated in quick succession, so that the razor disappears within seconds.

Fauna in clean sandy beaches cannot make permanent burrows because repeated wave action tends to collapse them; whereas in muddy beaches the burrow walls can be lined and strengthened with mud.

Surface clues

If nearly all the life of a sandy beach disappears out of sight as soon as the beach becomes accessible on foot, it would seem rather a dull place to visit at low tide. This is not so, because, even when the animals have burrowed away out of sight, many leave clues to their whereabouts on the beach surface, which are always fun to search out. A razor shell's burrow opening is like a keyhole, and even if this cannot be seen, a razor will react to vibrations set up as you walk by squirting a jet of water several metres up into the air. Many tube worms are easy to spot because the tops of their tubes project above the beach; while lugworms each produce a familiar worm cast, made by swallowing and passing out sand in their burrow.

PROJECTS

1 Using a metre-square frame, compare the density of lugworm casts at different tidal levels on the same beach and also on different beaches. Collect a handful of sand from the beach surface where the lugworms are most abundant. Shake it up with sea water in a screw-topped glass jar. How long does it take to settle? Repeat with sand from a beach where lugworms are least abundant. What do these experiments tell you about the type of beach preferred by lugworms?

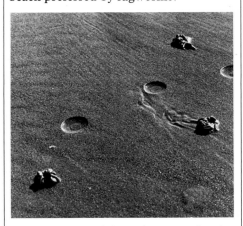

Lugworm casts and depressions exposed to the air during low tide.

2 Set up a wormery for observing the way in which a lugworm burrows. Place a piece of rubber tubing, positioned in a large U-shape, between two pieces of sheet glass each measuring 30 × 30 cm. Use three bulldog clips to press the glass sheets tightly against the rubber tubing. Fill the wormery with sand and top it up with sea water. Place a lugworm on top of the sand in the wormery. How long does it take to burrow?

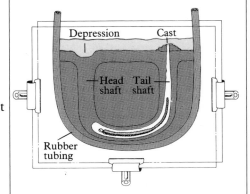

A lugworm *Arenicola marina* in a wormery showing the relative positions of the cast and depression.

3 Make a collection of empty shells found on the shore. Label each shell with the locality, habitat and date of collection. (The correct identification can always be added at a later date.)

123

SIEVERS AND SUCKERS

Many abundant inhabitants of sandy beaches live solely by feeding on food suspended in the water or on detritus which settles on the beach surface.

Sandy beaches lack the extensive seaweed blankets on which herbivores can browse on rocky shores. Apart from the odd clump of eel grass at the bottom of a shore, or the green seaweed *Enteromorpha* invading a freshwater seepage, the primary producers of a grazing sandy beach food chain are either microscopic diatoms living in fine deposits high in organic matter or free-living plankton in the sea. Many sandy beach inhabitants do not feed on conspicuous plants or animals, but on plankton which they filter from the water. On muddier shores, deposit feeders utilize detritus (the remains of plants or animals) instead. Energy flow takes place more slowly via a detritus food chain, since the dead organisms may have accumulated months before they are broken down as potential food.

Suspension feeders

Unlike herbivores or carnivores, which go in search of their food, suspension feeders tend to stay put and use their energy to collect and strain the particulate food from the water. They prefer coarse-grained deposits which do not clog up the sieve mechanism. Once the incoming tide submerges the beach, bivalves can open their valves, extend their short siphons up above the sand and begin to feed. A current of water is drawn in via the inhalant siphon by a mass of microscopic hair-like cilia beating on the gills. The food is strained off on to the gills, where it is mixed with a mucus secretion before being carried to the mouth. The wastes are excreted with the outgoing water current via the exhalant siphon.

Animals of sandy shores occur in zones, which are determined by the length of time they require to be submerged to feed, rather than by their tolerance to dry conditions. Like all intertidal filter-feeders, the edible cockle can feed only when submerged by the tide. Extensive cockle beds occur near low water level where the bivalves can feed for a large part of the tidal cycle. Where food supplies are plentiful, cockles can reach densities of over 1,000 per square metre. It is now known that not only is the feeding of

▼As soon as the sea covers the tube of the peacock worm *Sabella pavonina* it moves upwards (below) until all the fine tentacles project freely above the top. The tentacles can then open out like petals of a flower (right) in readiness to filter-feed.

▲Filter-feeding slipper limpets *Crepidula fornicata* compete for food with commercial oysters. They typically form mating chains, as here, with the smaller males attached above the larger females. As the males enlarge they change their sex.

▶The parchment tube worm *Chaetopterus variopedatus* removed from its tube to show the highly modified segments used for feeding. The largest segments in the centre of the body carry the three pairs of fans.

▼*Polymnia* is a terebellid worm which uses its elongated

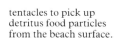

tentacles to pick up detritus food particles from the beach surface.

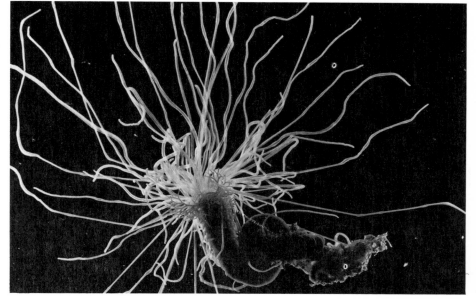

cockles linked with the tidal rhythm, but so is their digestion, which tends to be concentrated during the non-feeding period at low tide.

Razor shells (pp. 122–3) also filter-feed by drawing water in through the inhalant siphon extended up above the sand surface. Relatively few gastropod molluscs, however, feed on suspended matter, the slipper limpet being a highly efficient exception. This alien, which was accidentally introduced to Britain at the turn of the century with an importation of American oysters, has very long gill filaments which increase the area for trapping plankton. As well as competing for food with commercial oysters, the slipper limpet also competes for bottom space.

The sandy-coloured ends of the tubes made by polychaete worms which project up above the beach surface belie the beauty of the worms inside, most of which can be seen only when submerged by the sea. Even then, the peacock worm will withdraw immediately it senses a vibration or a shadow. The peacock worm is a fanworm which filter-feeds by trapping food on its delicate tentacles. Cilia carry the particles down towards the centre of the fan, where they are sorted. The smallest ones fall to the bottom of a gutter and are carried to the mouth, the middle-sized ones are used for tube-

building, and the large ones are rejected.

The way in which the parchment tube worm feeds can be seen only if it is transferred from its own leathery, U-shaped tube into a glass one. The first 16 segments in the body are highly modified for feeding. Segment 12, which has a pair of large flaps, secretes a mucus bag through which the food is strained. A water current is drawn in through the tube by the beating of three pairs of fans. Food is rolled up with the mucus into a ball and carried up to the mouth by cilia.

Deposit feeders

Animals which depend on a rain of detritus for their food also tend not to move around very much, but if need be they can burrow through the sand. There are four main methods used to collect deposits which settle on the beach surface. Firstly, the lugworm is an indiscriminate mud-swallower, feeding on the detritus mixed up with the beach particles and excreting the rest. Sand and detritus stick to the rough proboscis as it is pushed out, and are swallowed when it is inverted. So as to get enough nourishment, a lugworm has to 'feed' for a total of 5 to 8 hours per day. Lugworms live 20 to 30 cm down in an L-shaped burrow, which has an open shaft at the tail end. This is where the worm backs up to add to its cast. Blackened casts prove that the worm has been feeding within an anaerobic zone where sulphide bacteria have produced hydrogen sulphide. A few centimetres away from a lugworm cast, a shallow depression should be visible. This marks the top of the head shaft, where sand falls down as the lugworm feeds.

Deposit feeders which feed more selectively include the little amphipod *Corophium volutator* (pp. 112–13), which scrapes organic deposits from the surface around its burrow, and the bivalves which use their long, flexible inhalant siphons like miniature vacuum cleaners to suck up detritus. The fourth method of deposit feeding is utilized by some polychaete worms which extend highly extensible grooved tentacles over the beach surface. Cilia inside the grooves carry food particles from the beach surface to the mouth.

It is now known that several of the so-called detritus feeders do not digest the detritus, but feed instead on the bacterial film which grows on the particles. Just as woodland birds have differently shaped bills for exploiting different food sources, so the sandy beach fauna uses various methods for trapping its particulate food. For example, mucus sieves can strain much finer particles than combs made from the setae of arthropod limbs, and so the whole spectrum of the potential food supply is not utilized by all the fauna.

PROJECTS

1 Look for fanworm tubes projecting above a beach at low tide. Return just before the incoming tide reaches them. Stand quite still, making sure that you do not cast a shadow over them. How long after the tubes are covered by sea water does it take for the worms to emerge? **Warning**: Before staying down on the lower shore make sure you have someone with you who knows the beach well, so that you can safely return to the upper shore without danger of being cut off by the incoming tide.

2 Find a live cockle. Put it in a clear sandwich box with a layer of sand topped up with sea water. Make a suspension of brewer's yeast so that it looks milky. Using a pipette, squirt some above the open cockle siphons. Watch what happens.

An edible cockle *Cerastoderma edule* feeding under water with extended siphons.

3 Place a live tellin shell in a container with sand and sea water. Add a small amount of dental scale detector (available from chemists) to the sand surface to tint the sand bacteria. Watch how the tellin hoovers up the coloured particles with its inhalant siphon.

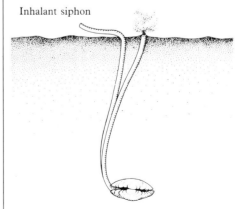

Exhalant siphon

Inhalant siphon

A Baltic tellin *Macoma balthica* buried beneath sand, showing siphons extending up above the beach surface.

A HOSTILE ENVIRONMENT

Sand dunes provide one of the most unstable and inhospitable places to live in, where drought and sudden fluctuations in temperature occur regularly. Yet they support a great variety of plant and animal life.

The golden hills of sand that everyone enjoys on a summer's day impose some of the toughest conditions on living things. Fates such as being buried beneath mounds of sand, exposed to the fierce rays of the sun in a soil that has no moisture in it, or drenched with salt spray, are regular occurrences in the dunes near the sea. Conditions change rapidly, from the expanse of shore where the sand is deposited, to the highest and oldest dunes that are furthest inland. Wind is the chief operator in the movement of sand, but as soon as the blown sand meets an obstruction such as an older dune or a plant tussock it piles up on the lee side of it. The relationship between the forming dunes and the plants that colonize them is not haphazard, but results in a series of wave-like crests and hollows that increase in size as they get further from the shore. Although they appear quite fixed to us the younger dunes move inland at the rate of up to 7 metres a year, or even more, depending on conditions, and take around 50 years to reach their maximum height and become fixed dunes. This resulting variation in topography and aspect gives a great variety of microclimates in which moisture and temperature conditions as well as sand movement restrict plants and animals to particular situations in the dunes. Despite the hard living conditions, the animal and plant life of the dunes is particularly rich and varied.

Living conditions

Sand is made of fairly large particles that allow water to run between them rapidly, so rain never remains long in the upper parts of the dune. Where there are large amounts of sand exposed in the mobile dunes, evaporation can rapidly reduce the value of a shower on a hot day. In the older slacks, or hollows, water may accumulate, giving a freshwater environment that contrasts with the drier dune tops. Although temperatures do not often fall below

◀Sea holly *Eryngium maritimum* and ▶sea bindweed *Calystegia soldanella* are unusual plants of the mobile dunes that have adapted to sand dune conditions, the sea holly with tough, leathery leaves and a deep root system, and the sea bindweed with an extensive, creeping stem which is protected by the upper layers of sand and leaves that remain close to the sand surface where humidity is higher.

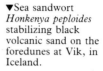

▼Sea sandwort *Honkenya peploides* stabilizing black volcanic sand on the foredunes at Vik, in Iceland.

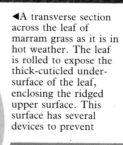

◀A transverse section across the leaf of marram grass as it is in hot weather. The leaf is rolled to expose the thick-cuticled under-surface of the leaf, enclosing the ridged upper surface. This surface has several devices to prevent water loss: the stomata are at the bottom of the furrows where little water vapour can escape, and the hairs prevent any air movement that would carry water vapour away from the inner surface.

▶Mobile dunes colonized by marram grass *Ammophila arenaria* in north Cornwall. Where the marram grass has been destroyed by trampling, blown sand has provided a space for another colonizer, the sea couch *Agropyron repens*, which is more typical of the foredunes.

freezing on the coast, very rapid fluctuations in temperature of up to 50°C occur between day and night. Once there is a continuous plant cover these fluctuations in temperature and moisture content are considerably reduced. With increasing plant cover there is also an increase in organic matter, but the mineral content of the sand is controlled by the nature of the material that forms the sand. If there is a considerable shell component, the dunes will be more alkaline, and particularly rich in species.

Adaptations to growing in sand

One of the most striking plants of the mobile dune system is the marram grass. This remarkable plant can be buried under metres of sand and grow straight up through it to form another tussock of wiry green leaves at the surface. The most impressive thing about it is the root system, often visible in eroded dunes. It has a vast network of roots and underground stems extending throughout the dune. In dry conditions the leaves appear dark green and rounded, but in wet weather they appear grey-green, exposing an upper, flat, ridged surface. A thin section of the leaf under a x 20 lens or a microscope reveals that the dark outer surface of the leaf has a thick cuticle and no breathing cells (stomata), while the inner or upper surface has fine, bristle-like hairs and breathing cells situated at the bottom of the ridges. These are all devices for restricting water loss from the leaf surface in dry conditions.

The foredunes are colonized by several fleshy-leaved plants which are adapted to the salt water spray that they regularly receive (p. 107). Sea purslane and sea saltwort are both widespread, low-growing plants with smooth, hairless, fleshy leaves that readily shake off the sand grains blown on to them.

They produce abundant, large, heavy seeds that either drop in situ or are rolled over the sand by the sea and thrown up on the strandline. In the older dunes, where wind-blown sand is less frequent, many of the plants grow a dense felting of hairs to reduce water loss, or they may grow in a compact rosette. Buck's horn plantain does both. Some common plants, such as silverweed, have a form specially adapted to sand dunes that is densely felted with hairs.

Adaptations to living in sand

Although animal life can move out of locally uncomfortable conditions, there is still a strong association between the plant microhabitats and the fauna. In a study on sand dunes in south-west Wales, 185 species of arthropod were caught from 9 different groups that included spiders, mites, ants, flies, woodlice, springtails, beetles and moths. Although many of these species are found in other habitats, they were found occupying particular niches in the sand dune system. For example, the red ant preferred the fore and mobile dunes, and the black ant preferred the fixed dunes. Spiders are very abundant in the sand dunes and were also restricted to areas of the dune system, but often changed these seasonally. Snails seem a very strange group to find in the sand dunes, yet they are often brightly coloured and conspicuous in the mobile and fixed dunes. They have a special way of avoiding desiccation; in dry weather they seal up their shells with a papery membrane to avoid evaporation, and aestivate (spend the summer, or dry weather, asleep) until it rains again. Their eggs are also resistant to drought and have been found to develop normally after 80 per cent of the water had been removed.

Other animals which may utilize sand dunes as their home include the sand lizard and the nocturnal natterjack toad, which will breed in pools in the dune slacks. During times of drought, the spadefoot toads will hide away in deep, sandy burrows. Evidence of temporary visitors to dunes can be seen by their tracks left behind in the bare sand.

By far the commonest mammal living in the dune system is the rabbit which maintains the close cropped turf that we associate with the fixed dunes. Without the rabbit the scrub would rapidly spread. In excessive numbers they may cause erosion of the fixed dunes with their burrows. But the greatest damage to the dune system is done by another abundant mammal – man. As he frequents the sand dunes in great numbers at the hottest time of the year, he may cause the destruction of the surface vegetation and so cause a blow-out in the autumn storms.

PROJECTS

1 Join sticky fly papers end to end on a wooden strip leaving plenty of space at the bottom to push into the sand. Make several of these, and place the strips so that they stand vertically facing the sea in different parts of the dune system, and in different wind conditions. Where does maximum sand movement take place? Does this relate to the size and mobility of the dunes?

2 Use glass jam jars as pitfall traps and sink these into the sand in different dune areas, being careful to rearrange the foliage over the opening. Leave overnight (but not any longer), and examine your catch in the morning. What groups have you caught? Which are you not likely to catch in these traps? When you release the animals on the sand in the day how do they behave?

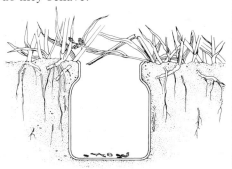

3 Record colour, number of bands, and location of specimens of the banded snail *Cepea nemoralis* throughout the dune system. Is there any correlation between either of these characters and the location of the snail? Look for remains of shells that have been predated in the mobile and fixed dunes, and record the same characters. Do your results suggest any relationship between colour, marking and predation in the banded snail?

The banded snail *Cepea nemoralis* in the sand dunes. The yellow snail has one band and the faded pink snail has three bands.

PREDATORS AND SCAVENGERS

After predatory animals have feasted and life generally dies and decays, scavengers do an invaluable job in mopping up the remains.

As in all food webs, the carnivores of sandy beaches are present in much smaller numbers than herbivores and detritus feeders lower down the web. Although relatively few of the animals which live in the sand itself are predatory, their numbers are boosted by the birds which descend to feed during low tide and the fish which swim up over the flats when the tide is in.

Lurkers

Good examples of cryptic coloration can be seen among predators which live in sand. Their bodies blend in so well with the sandy backcloth that their unsuspecting prey can rarely take avoiding action. Anyone who has collected live shrimps will know that the pinky brown colour appears after they have been boiled; in life, their bodies are a mottled sandy colour. During the day, shrimps bury themselves in the sand by using their legs to scoop out the sand from below and then their antennae to sweep sand over the body. When sandy flats are exposed to the air, the shrimps are buried below, but they can be collected by raking the sand surface. At night, shrimps emerge to feed on worms, small crustaceans and even young fish, as well as organic remains.

On a sunny day, the sand goby is easy to spot in the shallows low down the shore. If you stand quite still, the gobies will be seen darting to and fro, and maybe even investigating a bare foot or a shoe; but once a sand goby settles again, it blends in so perfectly with the bottom that it is almost impossible to spot. Sand gobies, which feed on small crustaceans, including shrimps, are taken by larger fish and by terns.

On shores where weever fish are known to occur, bare feet are definitely *not* advisable. The lesser weever is one of the few dangerous fish of European shores; the poisonous dorsal spines are capable of inflicting a painful wound in a foot and can even lead to a permanent injury if a secondary infection sets in. This predatory fish is well adapted for living in sand, using its pectoral fins to

dig a burrow, having eyes on the top of its head and an upturned gaping mouth for quickly snapping up its prey, which includes shrimps, sand eels and gobies.

Flatfish are also very well adapted for a bottom life, so much so that the body of the symmetrical, free-swimming planktonic larva undergoes a dramatic metamorphosis over a period of two to three weeks, before sinking to the bottom. Instead of having an eye on each side of the body, one eye migrates over to join the other, so that both come to lie on the upper side, giving a lop-sided appearance to the head. Flounders which move inshore up estuaries (pp. 114–15), as well as on to

sandy flats during high tide at night, lie on their right side and have their eyes on their upper left side; while plaice and sole, which lie on their left side, have both eyes on the right side. Either way, the flatfish sinks into the sand so that only its eyes protrude, ready to pounce on unsuspecting worms and molluscs which are crushed by powerful teeth.

Living in the sand, often remaining partly within their burrow, predatory polychaete worms shoot out a muscular proboscis, armed with jaws, to catch small crustaceans, worms and molluscs. The nemertine or ribbon worms also use a proboscis for feeding in a different way. Being long and thin, they capture

▲The head of a ragworm *Neanthes (Nereis) diversicolor* showing the extended proboscis ending in a pair of powerful jaws.

▶Netted dog whelks *Nassarius incrassatus* feeding on a dead crab *Carcinus maenas*.

▲A shrimp *Crangon crangon* beginning to burrow down into the sand, where it stays when the beach is exposed to the air.

During high tide at night, shrimps emerge to feed.

▲The bootlace worm *Lineus longissimus* lives beneath stones on muddy shores, using its proboscis to capture its prey.

▶*Actaeon tornatilis* moves through the sand using its foot like a miniature sand plough.

and hold their prey by twining themselves around it.

To feed on live bivalve molluscs, the carnivorous necklace shell has to bore its way through one of the shell valves. This it does by holding the shell securely with its muscular foot and etching it with an acid secretion before drilling a neat hole. Another predatory mollusc is the attractive little cuttle, which may get caught in shrimping nets in the summer. It can lighten or darken its body to blend in with the sand by expansion or contraction of light and dark pigment cells. Like its larger relatives the cuttlefish, squid and octopus, the little cuttle uses a beak in the centre of its suckered tentacles for crushing prey.

Active hunters

Silvery sand eels are most likely to be seen if they are washed up on to a sandy beach, since they mostly live offshore. Their smooth-sided, eel-like bodies, covered with tiny scales, offer little resistance to burrowing, which is done using a shovel-shaped lower jaw. Sand eels swim in large shoals at an oblique angle to the bottom, and when danger threatens they escape by darting down into the sandy sea bed. They feed on plankton as well as fish, and end up falling prey to seabirds (especially terns and puffins) and to larger fishes.

When terns are busy rearing their young, they can often be seen searching back and forth above inshore waters for sand eels which they delicately pluck from the sea surface. Although estuarine waders are covered in some detail on pages 110–11 mention must here be made of sanderlings, which return from the northern tundra to overwinter on European shores. The birds appear to work overtime as they move to and fro along the strandline in their search for sandhoppers.

Refuse eaters

The word 'scavenger' is commonly associated with vultures and other equally evil-looking animals; yet scavengers play a vital role in cleaning up rotting remains. Without them, the process of decay and recycling of nutrients would be much longer.

While the shore crab can be predatory, it will also scavenge for food. Sandhoppers may be small, but their sheer numbers ensure that much of the rotting strandline vegetation is consumed.

As netted dog whelks plough through sand, each holds its siphon upright, rather like a periscope, which is used to 'sense' the surroundings. As fresh supplies of sea water are drawn down inside the siphon, the scent of decaying food can be detected and the whelks rapidly home in on it. These whelks will converge on commercial whelk pots to feed on the crab bait.

Gaping cockle shells, with a tell-tale nick, are proof that an oystercatcher has been at work (p. 111); and herring gulls are opportunist feeders which swoop down at the edge of the sea wherever a starfish or a crab is stranded.

Nocturnal forays

Much more of the interaction of life on a sandy beach will be seen if a nocturnal visit coinciding with low tide is made; for then, as on rocky shores, the high humidity allows more animals to be out and about. A masked crab (pp. 122–3) can often be picked up by panning a torchlight beam along the water's edge.

PROJECTS

1 Walk along the bottom of a sandy beach at low tide, collecting all the empty banded wedge shells. What percentage of the shells have a neat hole bored by the carnivorous necklace shell? Is the hole always bored in the same position? Keep a bored shell to compare the shape and size of the hole with a mussel shell bored by a dog whelk (project 2 on p. 139).

A banded wedge shell *Donax vittatus* has been bored by the predatory necklace shell *Natica*.

2 Collect a live shrimp, gently transferring it to a sandwich box filled with sand and topped up with sea water. Watch the way it burrows into the sand and time how long it takes to bury its body completely.

3 Drag a dead crab or a fish along the bottom of a sandy beach where netted dog whelks are known to occur. How quickly do these whelks react to the presence of a food source? After leaving the bait, count how many whelks converge to feed on it. Repeat the above experiment at night. Are there any differences between the day and night recordings? If so, explain why this might be.

Netted dog whelks *Nassarius reticulatus*, previously buried in the sand, converge on a dead common whelk five minutes after it has been dropped into an aquarium.

LIFE AMONG THE PEBBLES

Shingle is produced by erosion from all land surfaces, but shingle beaches are found along exposed coasts where the sea has enough energy to carry and throw up a pebble ridge.

Standing on a shingle beach, out of reach of the waves, you can hear little but the sound of the sea, loaded with pebbles, as it thunders on to the steep shore, and then the roar of the rolling pebbles in the receding water. As you watch, the pebbles on the beach change their positions with each succeeding wave, the larger ones remaining higher up and the smaller ones being dragged down again. If the prevailing wind drives the waves at an angle the pebbles will also be sorted along the shore – a feature which is known as 'longshore drift'.

Although the pebbles between the tides are constantly rolled and polished, the pebbles above the high water level of neap tides are reached only by the spring and autumn equinoctial tides. The whole shingle ridge has usually been thrown up by storms, so the top part has very large pebbles and is relatively stable, neither wind nor rain being strong enough to displace it. This is known as a 'storm beach'. Unlike sand dunes, a shingle ridge rarely extends far inland but, since it may be higher than the land behind, it may alter the environment on the landward side by damming up fresh water or other deposits.

Where do the pebbles come from?

Many of the pebbles come from rocky headlands along the nearby coast, but along a shingle beach may be found a variety of rock types that are a considerable distance from their nearest geological location. Everything carried by water has its corners knocked off, not by the sea itself but by the other particles that the water carries, so even man-made materials, such as brick, concrete and glass, assume a pebble shape after being in the sea for some time.

Conditions for life

The most inhospitable place to live on a shingle beach is among the rolling pebbles on the foreshore, where the constant action of the waves and the lack of any organic material give no basis for life. Unlike the sandy shore there is rapid movement of water downwards through the shingle with each successive wave, and there are few crevices where the fauna can hide. Where the strandline adds organic matter to the shingle, and the pebbles are not subject to disturbance from spring to autumn, you may find annual plants such as orache as well as scavenging sandhoppers which will attract opportunist birds to feed along the strandline. Shrubby seablite is a woody perennial with deep roots which also occurs along the drift-line, where its buoyant seeds are dumped by the receding tide. The environmental conditions here are more favourable than in sand dunes, for, although water quickly passes through the pebbles, little evaporation can take place from the lower layers, and even on dry days

▲ A clutch of two eggs laid by a ringed plover is perfectly camouflaged on a bare shingle beach.

◀ Above the strandline on a shingle beach in Orkney, the attractive oysterplant *Mertensia maritima* flowers in midsummer.

▶ Above high water mark, encrusting orange lichens thrive on otherwise bare boulders, which also shelter a clump of shrubby seablite *Suaeda fruticosa*.

▶ The sheer size of the perennial seakale *Crambe maritima* makes it a striking shingle plant. Once sold in markets in the south of England, the plant is now very local.

pebbles are wet from condensation on their surfaces. Likewise temperature does not fluctuate as rapidly, both air and large pebbles being poor conductors of heat. Despite this, the surface appears quite barren, as the lack of fine soil particles and organic matter above the drift-line means that few plant species can colonize this habitat.

The colonizers

The colonization of a shingle ridge has

▲ By crouching down and keeping still, a ringed plover *Charadrius hiaticula* avoids detection on a vegetated shingle bank in Iceland.

the same sequence as many newly exposed surfaces of rock. Organic matter and soil particles being absent, the first colonizers on stable shingle are lichens and mosses which receive their nutrients from sea spray or rainfall. Once these have added to the nutrient material, other species that are tolerant of sea spray can germinate and anchor themselves between the pebbles. Among these are red fescue, sea campion and sea pea. These may form dense or loose mats over the surface of the shingle, but must keep a low profile as the top surface of the shingle ridge is exposed to the onshore winds. On northern shingle beaches may be found the beautiful oyster plant, so called because of the curious flavour of its leaves. Other perennials such as the yellow horned poppy and sea beet survive in a rosette or tussock during the winter season, but in the summer put up a show of flowers on a taller stalk. Once conditions are altered by these colonizers, terrestrial plants and animals rapidly move in from surrounding habitats.

The visitors

Many stable shingle ridges are valuable refuges to which various coastal birds return each spring to breed. Most of these birds feed elsewhere, but they require flattish coastal sites for rearing their young. Like some cliff-nesting birds (pp. 156–7), terns also breed in colonies, but they seek shingle or sand dunes instead of cliffs. Little terns prefer to nest widely spaced apart, but when they are forced to nest in artificially high concentrations in reserve areas, after loss of early clutches by disturbance elsewhere, they attract ground predators such as foxes and stoats. Aerial predators, such as the great skua, will also attack clutches exposed as a result of human disturbance, but they must first locate the nests from the air. Both the eggs and the chicks of these and other shingle-nesting birds, such as ringed plovers and oystercatchers, are so perfectly camouflaged with their surroundings that it is almost impossible to spot them, even at close range. There is no nesting material to draw attention to the nests; there is merely a hollow in the shingle, with one or two eggs that have their characteristic shape camouflaged by blotchy light and dark markings, in the same way as buildings were camouflaged during wartime.

Great care should be taken to avoid walking through any breeding seabird colonies, but where visiting is permitted in reserves, the presentation of fish during the courtship of little terns as well as the feeding of their chicks, may be safely watched from the paths by using binoculars.

STRANDLINES

Strandlines occur on all beaches, but they are most conspicuous on sandy shores. Here, a strange assortment of objects accumulates at the edge of the sea.

Flotsam and jetsam

The terms 'flotsam' and 'jetsam' are generally used to describe anything which is washed ashore, but this is not strictly correct under wreck law. If wreckage is cast ashore it is known as 'lagan'; if it is floating at sea it is 'flotsam'; and if it is cast overboard to lighten a ship it is known as 'jetsam'.

The position of the strandline varies with the high water level and is highest at extreme high water level of spring tides. During neap tides, when the high water mark occurs at progressively lower levels down the shore, a series of strandlines appears on the shore, all of which are pushed successively higher up the shore during the spring tide cycle. Although debris does get stranded on rocky shores, there are never any extensive lines such as those which occur in sweeping sandy bays. In among beached seaweed, a wide assortment of objects – both living and inanimate – can be found in the strandline which is accessible immediately after the tide has turned.

Floating objects, such as driftwood and cork, have always featured in strandlines; but nylon rope, plastic containers and pieces of polystyrene, as well as crude oil, all too frequently dominate the tidelines of today. In among this rubbish, much of which is non-biodegradable, an occasional gem can be spotted by the keen-eyed beachcomber. Stranded semi-precious stones include amber (fossilized resin from ancient conifer trees) and jet (a type of coal formed from a fossilized relative of the monkey puzzle tree). Whitby, in Yorkshire, on the north-east coast of England, was the European centre for the jet jewellery industry.

The conchologist will always scour a strandline in search of new or more perfect specimens of shells – especially those of offshore molluscs. Where shells dominate the strandline, the name 'shell beach' is typically used to describe the locality.

Other animal remains which get washed ashore include egg cases of molluscs, as well as those of skate and dogfish (known as mermaids' purses), cast crab shells, empty tests (known as

◄Like those of dogfish, horny egg cases of skate are known as mermaids' purses, and are not uncommon in the strandline.

▲As the tidal cycle moves from the spring tides into the neaps, a series of strandlines develops as each successive high water mark drops still lower down the shore.

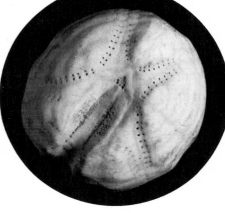

◄A tropical bean *Mucuna urens* being washed ashore on a Welsh beach.

▲A sea potato is the empty shell, or test, of the heart urchin *Echinocardium cordatum* which has lost all the spines. The paired holes are where the suckered tube feet emerged through the test.

◄Among the natural seaweed strandline on a shingle beach are numerous plastic objects, and an oiled guillemot.

►Creamish wrinkled egg capsules of the common whelk *Buccinum undatum* are common strandline objects. Sailors called them seawash balls, as they used them for washing.

sea potatoes) of the burrowing sea urchin and cuttlefish 'bones'.

Objects beached on European shores may originate from as far away as the tropics. Beans from the West Indies reach European coasts via the North Atlantic Drift current, which may also carry oceanic planktonic animals. On the west coast of Ireland, midwives gave tropical drift beans to women in labour, both to distract them and to bring them luck.

Oceanic drifters

On the south-west coasts in particular, unfamiliar animals periodically get beached during stormy weather and soon perish out of water. The Portuguese man-o'-war, the by-the-wind-sailor, goose barnacles and the violet sea snail all spend their lives floating on the surface waters far offshore, being carried to and fro in the currents. The mechanisms for keeping themselves buoyant vary from a gas-filled float (Portuguese man-o'-war) to a bubble raft (violet sea snail). Most goose barnacles hitch a ride on bottles or driftwood, but *Lepas fascicularis* is one which secretes its own float.

Wood borers

A closer look at driftwood may reveal clues to animals living inside. Holes and tunnels about 6 mm in diameter, which may be lined with a hard white cement, are the work of the shipworm, which is not, in fact, a worm, but a bivalve mollusc with a pair of very reduced shells. Greek and Roman wooden galleys, as well as Drake's *Golden Hind*, were all riddled with shipworm burrows. Today, shipworm damage is confined mainly to wooden piers and wharves; but the dykes, which are so vital to the existence of the Dutch lowlands, are also vulnerable to attack. The structure of a shipworm can be seen only when it is removed from its burrow. Most of the body is naked and worm-like, with the small pair of shells, highly modified for burrowing, at one end. Shipworms infest wood during the planktonic larval stage. At first, they enter the wood at right angles to the grain, but the main burrow always runs with the grain and one shipworm burrow never crosses another.

A much smaller but just as destructive wood borer is a crustacean known as the gribble, which resembles a miniature woodlouse. The 3–4mm-long gribble uses a pair of mandibles to bore through wood, particularly at the low water level of pier piles. In badly infested wood, gribbles are so numerous that as many as 300 to 400 can occur in 6.5 square cm.

Seaweed exploiters

Beached seaweeds dominate strandlines, especially after excessive wave action during winter gales has ripped up submerged sublittoral algae. This potassium-rich fertilizer is still exploited by farmers and horticulturalists, who now use forklifts in place of horses and carts to collect the mounds of rotting seaweed.

Man is not alone in utilizing seaweed. Sheep and cattle, particularly in Scotland, move down to feed on the shore. On the island of North Ronaldshay in the Orkneys, the sheep feed solely on seaweed, since they are confined to the shore by stone walls.

The most abundant animal which lives permanently in the strandline is the sandhopper, which burrows into sand beneath the seaweed. When the weed is turned over, a cloud of sandhoppers make random jumps a metre or so into the air, by suddenly straightening their bent bodies. Like all amphipods, sandhoppers have a body which is flattened from side to side. They scavenge – at night – on any rotting plant and animal remains. Unlike most shore animals, they have developed characteristics of a terrestrial way of life. There is no planktonic larval stage, since the female sandhopper carries her eggs and young around with her in a brood pouch.

The kelp or seaweed fly also lives and breeds in strandlines. Each developing embryo has its own air bubble inside the egg which ensures its ability to survive submergence by the sea for short periods. When the larval maggots hatch, they help to accelerate the decomposition of the seaweed. Sometimes, mass migrations of kelp flies move inland, where they are attracted into buildings by certain smells, one of which is trichlorethylene, a substance used for drycleaning.

Temporary visitors

Several birds, both wading and terrestrial, forage in strandlines for sandhoppers and kelp flies. Rock pipits, purple sandpipers and turnstones are among the shore birds which can often be seen busily working their way along the strandline. Starlings, pied wagtails and even robins will desert their inland haunts to feed on the shore; while at night, foxes and hedgehogs have been known to visit the shore to feed.

PROJECTS

1 Survey 1-metre lengths of strandlines on different beaches *on the same day.* Compare the abundance of living to inanimate objects on each beach. Which natural object is most abundant and why?

The sandhopper *Talitrus saltator* is one of the most numerous strandline animals.

2 Look for pieces of wood infested by wood-boring animals. Have they come from man-made structures? The wood will have to be split open to see the animals inside. What proportion of the wood you have seen on the beach is infested?

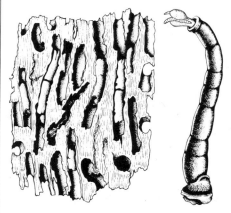

Strandline wood with neat parallel burrows, lined with a white cement, will be the work of the shipworm *Teredo norvegica*, which looks more like a worm than a bivalve mollusc.

3 Visit a beach as soon as possible after a *south-west* gale has been blowing, and look for surface drifters. Note generally the types of objects in the strandline and compare them with those present on the same beach a month later.

When working along a strandline *never* touch suspicious-looking objects or containers – they may contain hazardous chemicals or explosives

ROCKY SHORES

Rocky shores show more variation than any other coastal habitat, ranging from wave-cut platforms to steep cliffs, from smooth slopes to boulder-strewn beaches, from ledges to gullies. The geological structure of the rocks themselves shapes the finer shore contours such as crevices, rock pools and rocky overhangs.

Persistent wave action and oceanic swell severely limit the range of species which occur on exposed rocky shores, especially the seaweeds on the upper parts of the shore. On sheltered rocky shores, seaweeds provide a valuable protective blanket for animals to shelter from hot sun and dry wind, as well as a food source for browsing intertidal herbivores. Sheltered shores with a good seaweed cover and extensive hideouts such as crevices, rocky overhangs, boulders and rock pools support many different species totalling several thousands of seaweeds and animals. Although the number of species is much reduced on exposed rocky shores, the numbers of individuals of some species can be very high.

Only during low water of spring tides (LWST) is the lowermost zone of large brown oarweeds exposed to the air; and even then it is merely the upper fringe of a large zone which extends well down into the sub-littoral.

This lowermost part of a rocky shore with its extensive oarweed forest of *Laminaria digitata* and *Laminaria hyperborea* is exposed on only a few days each year during the biggest ranging spring tides.

ZONATION

A feature of all rocky shores, whether they be exposed or not, is the way in which the seaweeds and animals do not occur haphazardly all over the shore. They are found in distinct zones, with the dominant species describing each zone.

The vertical range of individual species of animals and seaweeds on the shore varies considerably, but none will tolerate living over the entire range from EHWS to ELWS. When the tide is in and the shore is bathed in sea water, intertidal life has no problems of water loss and the temperature of the surrounding water is uniform. The rising tide brings with it dissolved oxygen, carbon dioxide and nutrients. However, when the tide is out, the organisms have to face wide temperature fluctuations as well as the problem of drying out, both of which severely limit the level at which many species can survive on the shore.

If all shore profiles and rock strata were uniform, the zonation would be predictably monotonous. Fortunately this is not the case. Flat, smooth rocks which cannot hold water soon dry out; whereas cracks and crevices provide localized areas which retain moisture for longer and thereby allow animals to survive higher up the shore.

Causes of zonation

The shore zones are thus clearly related to the tidal levels, with the distribution of species linked with the relative amounts of exposure to air and submersion in sea water that each will tolerate. In addition, the amount of exposure to wind and swell, the extent of wave action, the shore profile and the way in which the shore is weathered to provide microhabitats all influence zonation. However, although the length of exposure time is crucial in determining the presence or absence of sharply zoned species, interaction between species and competition for rock space and for food, as well as predation, are also important factors. Mobile animals, such as winkles and topshells, may move around within their zone to feed, but if they are taken above or below their preferred zone, they will migrate back to it except at low water. The competition for space to live on the shore is so great that when an area of rock is completely cleared of all

Rocky shore zones as they relate to the tidal levels.

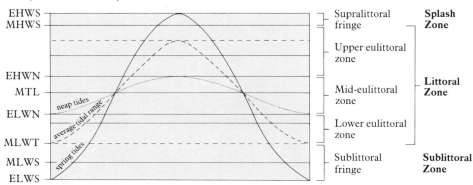

EHWS Extreme high water level of springs
MHWS Mean high water level of springs
EHWN Extreme high water level of neaps
MTL Mean tidal level

ELWN Extreme low water level of neaps
MLWT Mean low water level of tides
MLWS Mean low water level of springs
ELWS Extreme low water level of springs

◄With a mean spring tidal range of over 9 metres, the change of scene on a rocky coastline in Guernsey, during submergence at high tide and exposure at low tide, is very dramatic. At low tide, a sandy bay is exposed and the fringing seaweed band is visible in the shallows.

▶With a mean spring tidal range of under 3 metres, the seaweed zones on Icelandic shores are so compacted that on a single rock face the bottom of the upper zone of channelled wrack is separated from the top of the lowest zone of oarweeds by a mere 2 metres.

life a wide range of species recolonizes it; but only those best adapted will survive.

The main zones

A zone is a band on the shore in which certain species dominate. Three major biological zones are recognized on rocky shores: an upper black lichen zone; a middle barnacle zone; and a lower laminarian zone. A more precise way of describing the position of a particular species in relation to the tidal levels is to use the initial letters (see diagram opposite) of the various tidal zones.

Right at the top of the shore, the supralittoral fringe, which ranges from EHWS to MHWS, is covered by only extremely high tides. Below this, the upper eulittoral zone extends from MHWS to EHWN; the mid-eulittoral zone from EHWN to ELWN; the low eulittoral zone from ELWN to MLWT; and finally, the sublittoral fringe from MLWT to ELWS. The mid-eulittoral is the least stressful zone for its inhabitants, since it is covered by every high tide and exposed at every low tide.

Exposed shores

On exposed rocky shores, the upper zones are expanded by wave action. On the Atlantic coasts, which are constantly battered by waves and swell, the extent of the sea spray—marked by the black tar-like zone of the lichen *Verrucaria*—may extend vertically 20 metres above MHWS. However, in sheltered sea-water lochs and on far northern shores with a small tidal range, the lichen, barnacle and seaweed zones are all compressed within a vertical range of a mere 4 metres or less. The eulittoral zone on exposed shores is dominated by sessile filter-feeding barnacles and mussels. Seaweeds are lacking on exposed shores, except at LWST. The edible brown seaweed dabberlocks can withstand considerable

buffeting by waves and so it occurs on exposed shores near low water.

Sheltered shores

Because no two sheltered rocky shores are identical, there will be local variation between the relative abundance of the common species; but each one is most likely to occur at a specific tidal level on the shore. The zoning of the fucoid brown seaweeds is governed by their ability to withstand drying as well as their rate of growth. The first one to be encountered on a sheltered shore is channelled wrack in the upper eulittoral. Below this, bladder wrack and knotted wrack are less tolerant of drying, but grow rapidly in the mid-eulittoral.

Finally, the large oarweeds or laminarians survive in the sublittoral fringe where their pliable stalks or stipes allow them to move with the water rather than to resist it. Seaweeds are attached to rocks by a swollen base or holdfast. Laminarians have a large branching holdfast which is another adaptation for survival in turbulent water. The delicate red seaweeds appear low down the shore and are the last plants to penetrate the depths offshore, where the fading light limits, and eventually excludes, plant life.

Once seaweeds are established, they provide both shelter and food for many animals. The range of animal species increases as the time exposed to the air decreases, but the number of animals which live among the common intertidal seaweeds far exceeds the density of animals in the richest soil on land. In among channelled wrack, some 2,400 animals have been found living on one square metre of rock; whereas associated with bladder wrack as many as 50,000 animals can live in the same area, and over 200,000 per square metre may be associated with knotted wrack.

PROJECTS

1 Compare dog whelk shells from an exposed site with some from a sheltered shore. Using graph paper and a millimetre ruler, draw the outline and the shell opening (aperture) of the shells to scale. Which ones have the biggest aperture? Why should it be an advantage for these dog whelks to have a large aperture?

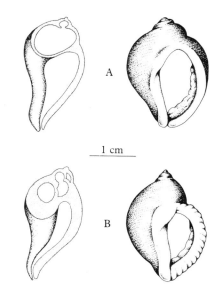

1 cm

Dog whelk shells (*Nucella lapillus*) from an exposed shore (A) and a sheltered shore (B). Note the enlarged shell aperture in A; and the thicker shell (an adaptation against predation) and correspondingly smaller aperture in B.

2 Mark individual edible winkles with a small blob of quick-drying modelling paint. Note their position on the shore in relation to some obvious landmarks. Return to the shore at weekly intervals to see if the winkles have remained in the same zone.

3 Look for a sharp upper line of seaweeds, barnacles or mussels fouling sea walls, pier supports or other man-made structures. Find out the time of high water and of low water by looking in a local paper or in tide tables sold at fishing tackle shops. Time how long the upper zone is covered by the sea and how long it is exposed to the air.

ON THE ROCKS

The most conspicuous animals on rocky shores are those which live a sedentary life out in the open, permanently attached to rocks. Most numerous are pale brown acorn barnacles and dark blue mussels.

When these fixed animals are exposed to the air, they cannot crawl away like crabs or winkles. Instead, they have to tolerate and survive high temperatures and desiccation on a sunny day; a freshwater shower during a rainfall; and low air temperatures in winter. When exposed, barnacles pull down their central upper plates and mussels close their paired shells. Only when submerged by sea water do they re-open and begin to feed on the fine particles suspended in the water. Barnacles use their hairy legs as a net to comb out the particles; mussels draw a water current inside their shells and strain off the particles, which are mixed with mucus before being swallowed. Inside the mussels, orange gills covered with masses of tiny hairs (cilia) set up the water currents by beating in unison.

Balanus balanoides is a common intertidal acorn barnacle which lives for 3 to 5 years depending on the tidal level where it settles. Those barnacles which settle low down on the shore stay submerged for a large proportion of the tidal cycle and so they can feed for a longer period each day. They therefore grow quickly and are able to breed after only 1 year, but they die after 3 years. Barnacles which settle higher up the shore are submerged for a shorter period during each high tide and so have less time for feeding. They take 2 years to reach breeding size, but have a life span of 5 years.

Both acorn barnacles and mussels produce vast numbers of tiny larvae. As these larvae feed and grow, freely floating as plankton for several weeks, they are widely dispersed by water currents before they settle and change into adults. The barnacle cypris larvae are attracted to settle close to other barnacles by the chemicals they secrete. Once a barnacle cements itself to the rock it is fixed for life, and so it must find its potential mate before it settles. Although mussels do not mate, they do need to be concentrated together to increase the chance of the eggs being fertilized in the surrounding water.

▲Every portion of this eroded limestone shore is covered with inter-tidal invertebrates– dog whelks *Nucella lapillus*, mussels *Mytilus edulis*, limpets *Patella vulgata* and barnacles.

▶A pair of mussels *Mytilus edulis* filter-feeding under water. In between each pair of valves can be seen a pair of siphons.

▶When it is submerged by the sea, this edible winkle *Littorina littorea* crawls over rocks using the broad muscular foot.

cannot grow outwards if they are densely packed, as pressure all round forces them to grow upwards instead, forming barnacle mounds. Since their bases then cover only a small area, they are easily dislodged in rough seas. The superabundance of barnacles or mussels tends to occur in alternating cycles, with one or other dominating at any one time.

Although barnacles and mussels have hard shells they are still preyed upon by other animals from which they cannot take evasive action. Dog whelks prey on barnacles by prizing open the upper plates, and on mussels by boring a hole through one shell. Blennies are shore fish which have rasping teeth for munching barnacles. A starfish is able to pull the shells of a mussel just far enough apart for it to push its stomach through the tiny gap and start to digest away the soft body tissues.

Limpets, winkles, dog whelks and topshells are mobile, hard-shelled molluscs which live on rocky shores. When exposed to the air during the day, limpets stop feeding and clamp their shells down on to the rock. Each limpet has a precise 'homing' position to which it returns after its feeding forays by crawling on its muscular foot. On soft sandstone rocks, a limpet grinds a circular scar by rotating its shell until it makes a good fit with the rock; on hard quartz rocks the shell itself is ground down to fit the rock contours. The large foot acts like a suction disc, clamping the limpet to the rock with a grip that can be quickly tightened if the limpet is knocked by a wave or would-be predator.

Limpets feed by rasping off microscopic algal growths from the rock surface, using a long, toothed ribbon called a radula, which soon gets worn down and is constantly being renewed. During its first year of life, for each gram of weight, a limpet must eat 75 square cm of encrusting algae. Limpets can be very numerous (as many as 182 have been recorded in a square metre), and then their constant browsing on young seaweed sporlings prevents the growth of larger plants. The way in which limpets maintain bare rocks could be seen very clearly in 1967 after the *Torrey Canyon* disaster. On shores in south-west Britain, where all the limpets were killed by detergent, the brown rocks turned green as sea lettuce survived and flourished.

The profile of a limpet shell is influenced by the amount of its exposure. Limpets living in rock pools have low shells with broad bases, whereas those on exposed rocks develop high shells with narrow bases. The exposed limpets have repeatedly to clamp their shells on to the rock. This also pulls in the marginal tissues, thereby restricting the shell growth.

◄Young acorn barnacles *Balanus balanoides* were attracted to settle close to adult barnacles by a chemical stimulus.

▲Limpets *Patella vulgata* exposed to the air during low tide, showing rock bared by their browsing activity as well as some old limpet scars.

Mussels are anchored to rocks by byssal threads which they secrete. Newly settled mussels and other bivalve (two-shelled) molluscs are know as 'spat'. In years when there is abundant food heavy spatfalls of mussels or settlements of barnacles occur, completely smothering the rock surfaces. If there is little available space on the rock, this will limit larval settlement. Even if high numbers of larvae do settle, lack of space and food will mean that as they grow not all will survive to reach maturity. Also, individual barnacles

PROJECTS

1 Find a small stone or shell to which live barnacles are attached. (Dead barnacles have an obvious hole in the centre of their outer plates.) Transfer the stone or shell to a transparent sandwich box filled with sea water, or a shallow pool, and observe the way the centre plates open out so that the barnacle can comb the water with its limbs. Using a second-hand on a watch, time the feeding rate. Take the temperature of the water. Transfer the stone or shell to a warmer or a colder pool. Note what happens to the feeding rate.

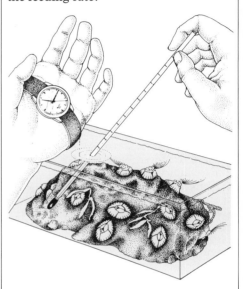

2 Look for empty mussel shells which have a neat hole bored through them. This is the work of a dog whelk. Is the hole always in the same part of the shell? Is this a thick or thin part of the shell?

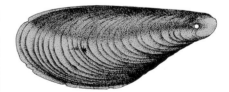

Mussel shell bored by dog whelk

3 Search for feeding trails of herbivorous molluscs. Measure, draw and compare them. Are they more obvious at certain times of the year?

Feeding trail made by common limpet

KEEPING MOIST

When exposed to the air by the ebbing tide, intertidal life faces the danger of drying out unless it is specially adapted or can take evasive action.

As the tide ebbs and flows over a rocky shore, life between the tides has to face wide daily fluctuations in the air temperature. High temperatures and strong winds reduce the relative humidity and hence accelerate water loss by evaporation from marine animals and seaweeds exposed to the air.

Moving elsewhere

One obvious solution for avoiding desiccation would seem to be simply to crawl off the bare rock into a rock pool; but as explained on pages 150–1, pools are by no means congenial, cosy habitats. In fact, most mobile animals seek shelter and shade during low tide by crawling into crevices or beneath boulders and seaweeds. These animals include crabs, sea slugs, sea urchins and fish. Small crabs lose water by evaporation more rapidly than larger crabs, because their surface area is large compared with their body volume.

Seaweed adaptation

Seaweeds—especially the high level ones—can lose over half of their total water content, because it is lost from the cell walls without damage to the cell interior. Channelled wrack extends higher up the shore than any other brown seaweed. The short growth form of this wrack, its inrolled margins and high oil content, are all adaptations which help this plant to resist desiccation for long periods of exposure. The young stages of knotted wrack are much more vulnerable to desiccation than tougher, older plants, which help to nurse the young sporlings. Knotted wrack will also germinate beneath a cover of fucoid wracks, eventually completely outgrowing and replacing them.

Waterproofing

Whereas land insects cover themselves with a waxy macintosh, some marine worms and many marine molluscs and crustaceans use their hard outer shells as a barrier against water loss. When sedentary mussels and barnacles are exposed to the air, they protect the soft inner tissues by closing their shells;

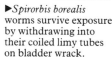

▲On the walls of a cave low down on a shore, encrusting animals, such as this elephant's ear sponge *Pachymatisma johnstonia* flourish in the damp environment.

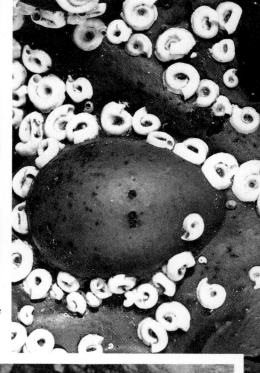

▶*Spirorbis borealis* worms survive exposure by withdrawing into their coiled limy tubes on bladder wrack.

▶The shell margins of a common limpet *Patella vulgata* have grown to match the rock contours and thereby make a perfect fit.

▼Part of a rock face exposed at low tide shows limpets, winkles and contracted beadlet sea anemones *Actinia equina*.

while limpets pull their conical shells snugly on to the rock. Since winkles and dog whelks are unable to clamp their shells tightly on to rocks like limpets, they crawl into crevices or shady overhangs. They can reduce water loss still further by completely withdrawing into their shells and sealing the opening with a horny operculum.

Edible winkles are often stranded attached to the sides of boulders by a thin mucus film. On dry days, the winkles are stimulated to secrete mucus which glues the shell margin to the rock as it dries. Once the mucus hardens, the foot is withdrawn and the operculum seals the opening. Molluscs which are exposed to the air for longer periods tend to have thicker shells than individuals of the same species which live lower down the shore. Polychaete worms are very much a feature of sandy shores, but there are also some tube-dwelling worms which live on rocky shores. The hard, limy tubes secreted by serpulid worms function in the same way as mollusc shells. The tiny, coiled, snail-like *Spirorbis* tubes can be found on brown wracks, while the irregular keelworm tubes occur on stones and empty shells. When the worms are exposed to the air, the feathery gills are withdrawn into the open end of the tube, which is also plugged with an operculum.

When to be out and about

Human beings flock to the shore to expose themselves to the full sun; shore animals do quite the reverse. If they are not specially adapted for survival in full sun, they must beat a hasty retreat. This means that most of the life on a rocky shore will be out of sight on a summer's day, whereas a visit to the same shore exposed on a cool, misty or rainy day, would reveal much activity out in the open. At these times, the shore does not dry out quickly and so the relative humidity remains high (85–90%). Instead of being tightly clamped to the rocks, limpets crawl around with their shells raised clear of the rocks, browsing on small algae. Flat-topped winkles remain on top of the brown wracks, crawling with their tentacles outstretched, and the attractive green leafworms crawl over the barnacle-encrusted rocks.

Even more activity is evident at night, when shore crabs emerge from their hideouts to feed on the open rock surface; prawns and fish emerge from the fringing cover of their pools to feed in the open water, and sea slaters crawl out from crevices at the top of the shore.

Soft-bodied animals

There are a few soft-bodied animals which are very resistant to desiccation, even when they are exposed by day. The beadlet sea anemone and the club-headed hydroid both secrete a mucus macintosh which helps to resist drying. The anemones are by no means an uncommon sight on exposed rocks at mean tide level, with their tentacles withdrawn inside the column which contracts down into a jelly-like blob.

Crevice fauna

The problem of desiccation is much less in a crack or crevice than on the open rock surface, and so animals are able to live higher up the shore in crevices. Similarly, sublittoral species can extend up into the lower fringes of a rocky shore, if there are rocky overhangs, deep gullies or shady walls of a cave—all places where the relative humidity remains high. These sites, which are accessible on foot only at the lowest spring tides, are often a multi-coloured mosaic of encrusting sponges, hydroids, sea anemones, sea mats and sea squirts.

The spaces between barnacles are miniature crevices where tiny marine crustaceans and worms, as well as some terrestrial insects, seek shelter. The insects have the reverse problem: to survive, they must remain dry. When submerged, they crawl into small pockets of trapped air, keeping the water at bay with their water-repellent bodies.

PROJECTS

1 Visit a rocky shore on a hot, sunny day, and count the number of flat-topped winkles you find crawling on top of brown wracks within a metre-square frame. Make a sketch of some obvious landmarks on this part of the shore, so you will be able to find the same spot again. Return to the same shore on a cold, overcast day and repeat the count. Are the two counts very different?

A flat-topped winkle *Littorina littoralis* crawls with outstretched tentacles over bladder wrack exposed to the air on an overcast day.

2 Compare different rates of water loss by evaporation. Take four identical glasses and put the same amount of water in each one, marking the water level in each with a black felt pen. Put the first in a cold place. Put the second in a warm place (airing cupboard). Leave the third at room temperature, but cover the top with a piece of cardboard. Leave the fourth also at room temperature, but first pour a thin layer of cooking oil on top of the water.

Leave all the glasses for a week, then measure either the depth of water loss below the initial black mark, or the amount of water remaining in each glass by using a graduated measuring cylinder. Which glass lost most water? Which one lost virtually no water? What does this experiment tell you about life on the shore?

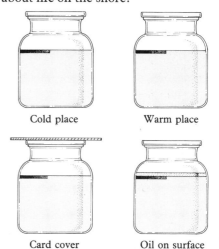

Cold place   Warm place

Card cover   Oil on surface
(room temperature) (room temperature)

CRABS

Although not confined to rocky shores, crabs are very much a feature of them, emerging at night to scavenge or prey on their food. If crabs are exposed to the air during a sunny day, they soon die from desiccation.

Getting around

On a worldwide scale, crabs are a very diverse group; even in Europe, the shape, colour and size are highly variable. Most crabs have five pairs of walking legs, the first pair of which ends in pincers.

Crabs can be grouped by the way in which they move around. Most rocky shore crabs move by walking or crawling: notably the shore crab, the porcelain crabs and the long-legged spider crabs. The edible crab, which is one of the signs of the zodiac, is a slow-moving crab, with a thick outer skeleton and a pair of massive pincers, which lives low down the shore and well into the sublittoral.

Another way by which some crabs move is by swimming through the water. Crabs belonging to the family Portunidae are active predators which swim rapidly after their prey, using the last two flattened sections of their hind legs as swimming paddles. The velvet swimming crab is a spectacular—and ferocious—shore crab which sports a pair of bright red eyes and deep purple body markings. French fishermen aptly refer to it as *le crabe enragé*. Only a few crabs are adapted for burrowing into sand (pp. 122–3).

Feeding

Crabs feed by scavenging on dead remains and by preying on live food. Porcelain and hermit crabs stir up muddy detritus, scooping it up to the mouthparts, which filter off the fine particles. Predatory crabs, such as the shore crab, the edible crab and the velvet swimming crab, have strong pincers for seizing and crushing their prey—usually molluscs. The tough claws of the box crab persistently chip away at gastropod shells to get at the flesh inside. Edible crabs will excavate pits up to 20 cm deep to reach burrowing bivalves.

Omnivorous spider crabs have elongated pincers which they use to pick up pieces of seaweed and small crustaceans—they can even collect food from inside small crevices. The wrinkled crab gathers red and green

▶As the velvet swimming crab *Macropipus puber* rests in shallow water, the flattened swimming paddles are clearly seen. When this crab is disturbed, it will rise up on its hind legs waving outstretched pincers.

▼Underside of female velvet swimming crab *Macropipus puber* showing egg mass held by the abdominal flap.

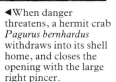

◀When danger threatens, a hermit crab *Pagurus bernhardus* withdraws into its shell home, and closes the opening with the large right pincer.

▲Like all spider crabs, *Macropodia rostrata* has long thin legs and a triangular body. The red and green seaweeds are effective camouflage on a weedy bottom.

▲The spiny spider crab *Maia squinado* adopts a defensive attitude by interlocking its legs. This crab is eaten by man in the Channel Islands and in mainland Europe.

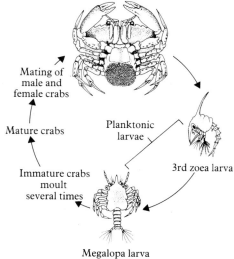

Gravid female

Mating of male and female crabs

Mature crabs

Planktonic larvae

3rd zoea larva

Immature crabs moult several times

Megalopa larva

Life cycle of the shore crab *Carcinus maenas*

seaweeds by plucking them off rocks with its pincers.

Moulting

For most animals, growth is a gradual process; but the hard outer skeleton makes this impossible for crabs. The carapace functions like a plate of armour against attack by predators, but it cannot gradually expand to accommodate a growing crab. Instead, it is periodically shed, after a new larger carapace has been laid down beneath it. As the old carapace splits, the crab backs out from it, leaving behind a complete skeleton. The new carapace is very soft at first, so these 'soft' crabs are vulnerable to attack by predators and therefore tend to hide away, but they are much sought after by sea anglers for bait. The frequency at which a crab moults depends on its age. During the first year of life, a shore crab may moult twelve times, and thereafter only once or twice a year. When hermit crabs moult they also have to find a larger shell in which to take refuge.

Autotomy

Crabs have a means of escape from attackers akin to lizards' shedding their tails. They are able to shed, or autotomize, their limbs along a predetermined breakage plane. When pressure is put on to a leg, either by a predator or by a stone falling on to it, an autotomizer muscle inside the leg contracts, suddenly causing the leg to snap off close to the carapace. This reflex action is an important survival adaptation, but a shed limb can be replaced only when a crab next moults. Loss of body fluids is prevented by an internal partition covering the leg stump. If a crab loses a walking leg, it is not so serious as losing a pincer, which is used for feeding.

Breeding

The very elongated pincers of the male masked crab readily distinguish it from a female. A useful way of sexing crabs which do not show such a marked sexual dimorphism is to look at the shape of the abdominal flap beneath the body. Female crabs tend to have a broad flap for carrying their eggs. Before the eggs are laid, however, crabs must mate. Shore crabs, like several others, mate as individual pairs, with their undersides pressed together, whereas spiny spider crabs congregate in shallow water offshore in large conical heaps during the summer. The larger, mature male crabs cluster on the sides and top of the heap, protecting both male and female crabs which are moulting inside. After moulting, the soft-shelled female crabs mate with the hard-shelled mature males. These females, in common with other species in which the females mate with soft shells, do not lay their eggs for some months—as many as 14 months in the case of the edible crab.

When a female crab is carrying her eggs, she is described as being 'in berry'. The eggs hatch into planktonic larvae, which pass through a series of moults until they change into tiny replicas of the adult.

Disguise

Crabs tend not to be exhibitionists; instead they usually blend in with their surroundings. The spider crabs, in particular, are masters of disguise. Tubercles and hooks on the carapace encourage seaweeds to settle and grow, assisted by the crabs' 'planting' pieces of seaweeds on their own backs. Hydroids, sea mats and sea squirts will settle on the backs of sublittoral spider crabs, thereby helping to camouflage them. The sponge crab walks around carrying a sponge on its back, held in place by the last two pairs of walking legs, while the hairy crab traps silt in its hairy body, which helps it to blend in with a muddy bottom. Although crabs blend in quite well with their surroundings, they are still frequently eaten by predatory octopuses, fish, birds and mammals.

1 Search for shore crabs and sex them by looking at the shape of the abdominal flaps. In female crabs, the abdomen has 7 segments and is rounded; in the male it is more pointed, and since the middle 3 segments are fused there appear to be only 5 segments. The safest way to hold a crab is by the back of the carapace, where the pincers cannot reach. Are the males and females present in equal numbers?

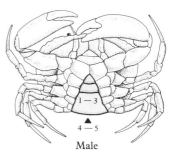

Male

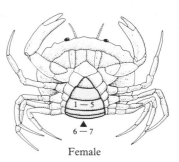

Female

Underside of male and female shore crabs *Carcinus maenas*, showing different-sized abdominal flaps.

2 Go down on to a rocky shore during low tide, at night, with a torch or a spirit lamp and observe crabs feeding on the shore. Choose a shore which you have visited by day, so that you know the terrain well. (Once you have spotted a crab do not keep the light shining directly on it for too long.) How many species of crab are there? Do you find the same number on the upper shore as on the lower shore, or are they fairly localized? What are they feeding on? Does one kind of crab appear to prefer one kind of food?

A shore crab *Carcinus maenas* emerges at night from its hideout to feed on a limpet on bare rock.

BREEDING FOR SURVIVAL

For a species to survive, it must produce offspring. Among rocky shore invertebrates a wide range of breeding methods are utilized, including asexual and sexual reproduction, egg-laying and viviparity.

Types of reproduction

A few rocky shore animals, including some hydroids and sea anemones, reproduce asexually, without ever meeting a mate. They simply bud off small replicas of themselves, which are genetically identical.

The majority of shore animals, however, breed as a result of the fertilization of eggs by sperm. They may discharge the eggs or fertilized larvae directly into the sea, or they may lay the eggs on stones, shells or seaweeds; or they may brood their young inside them. Barnacles and sea slugs are hermaphrodite animals, each of which contain functional ovaries and testes. When a pair of their own kind meet, they are able to fertilize each other's eggs, so that two batches are laid.

The sea hare is a mollusc with the remnants of an internal shell, and it forms a link between the shelled molluscs and the true sea slugs. In the spring, the purple fleshy molluscs move inshore to spawn by forming a mating chain of up to six individuals. Each hermaphrodite sea hare functions as a male to the animal in front and as a female to the one behind it. The pink spawn ribbons laid among seaweeds resemble a mass of untidy pink wool.

Most shore animals are dioecious (that is, they have separate sexes), but they by no means all mate to ensure cross-fertilization of their eggs. Many simply discharge their eggs and sperm into the sea where they are fertilized externally. This method is clearly rather haphazard, and so vast numbers of tiny eggs are produced to allow for a large percentage of wastage.

Synchronous spawning

Since any type of reproduction demands an output of energy, it is important that the best possible chance of success is ensured. To this end, some animals form communal breeding piles.

▲A pair of acorn barnacles *Balanus crenatus* with their feeding limbs extended under water. One barnacle (right) is mating with its neighbour by means of an extensible penis.

▲A male rock goby *Gobius paganellus* stands guard over its egg mass.

▶The male worm pipefish *Nerophis lumbriciformis* carries a double row of eggs on his belly until they are hatched.

▶Dogfish *Scyliorhinus caniculus* egg cases, known as mermaids' purses, attached to brown seaweed by the long tendrils. Each sizeable egg contains a large yolk supply.

Spiny spider crabs cluster in huge heaps offshore (pp. 142–3); while dog whelks congregate in rock crevices to lay their yellow, flask-shaped capsules.

Many other animals which discharge their sex cells into the sea increase the likelihood of successful fertilization by synchronizing the release of the eggs and sperm with a certain phase of the moon. Some species spawn at full moon, others at new moon or the first or last quarter. Examples of animals which concentrate their spawning effort in this way include bristle worms, winkles, mussels, oysters, sea urchins and starfish.

Although the king ragworm lives in sand flats, it is none the less a particularly interesting example of a moon phase spawner. At full moon in April, worms up to 70 cm long swarm in the Danish Kattegat Sea. Their bodies are packed with eggs or sperm, and so many worms congregate in shallow water that they fill fishermen's nets, while both gulls and plaice gorge themselves with the plentiful food. After spawning, the worms die, and the fertilized eggs hatch into planktonic larvae. Many shore animals breed in the spring so as to coincide with the large spring plankton blooms on which the larvae feed.

Fate of the eggs

The best proof of recent breeding activity by shore animals is seen as eggs and spawn which are laid on the shore itself. Many molluscs, in particular, lay conspicuous egg masses. These include cowries and whelks which lay capsules; sea slugs which lay spawn ribbons; and

◄A cluster of dog whelks *Nucella lapillus* with their grain-shaped egg capsules.

▼A pair of sea slugs – known as sea lemons – *Archidoris pseudoargus* with their spawn ribbons each containing masses of minute eggs exposed to the air on the underside of a boulder.

the common octopus which lays large white egg clusters containing 30,000–50,000 eggs. The attractive bright green leafworm lays minute green eggs in gelatinous sacs, while the lesser spotted dogfish lays a few large, horny, brown capsules which are attached to seaweeds by tendrils extending from all four corners.

Care of the eggs is the exception rather than the rule. However, some animals carry them around, while others brood over them until they hatch. Female prawns, crabs and lobsters and all the pipefish males carry the eggs cemented to their abdomens; while the males of the two European seahorses carry them in special brood pouches. Several other fish, notably the butterfish and the male shanny, stand guard over their eggs

until hatching takes place. The production of planktonic larvae is essential for the dispersal of sedentary shore animals. This dispersal phase, in general, ensures the recolonization of a badly polluted beach. Since these microscopic larvae are eaten quite indiscriminately by filter-feeding animals, huge numbers of eggs are produced to allow for this high wastage; for example, three million are laid by the edible crab. Eggs are eaten in other ways. Dog whelks lay several hundred eggs in each of their capsules, most of which are infertile and provide food for the 10–12 fertile eggs. *Calma glaucoides* is a sea slug which feeds almost exclusively on the eggs of intertidal fish, including blennies and gobies. As many as 60 sea slugs have been found on one egg mass. The fish known as the lumpsucker lays its egg clusters on rocks, where they are preyed upon both by rats and birds when they are exposed to the air, and by fish when they are covered by the sea.

Dogfish and skate put their reproductive effort into producing relatively few large, yolky eggs. This ensures that the developing embryos are well nourished and allows them to grow to as much as 10 cm long before they hatch out nine months later. This prolonged development, compared with the few days it takes planktonic larvae to hatch from their eggs, means that infant mortality in dogfish is very low.

Winkle variation
It is not unexpected that the life history and the kind of spawn produced by quite unrelated species should differ; but even between the four species of intertidal winkles there is a marked difference. This variation is linked with their ecology, especially the tidal level at which they live. The small winkle lives at the extreme top of the shore. It spawns during the winter (September to April), when the tidal level extends higher up the shore during rough weather. Planktonic egg capsules are then released during the spring tides. The edible winkle also produces planktonic egg capsules but, since it lives on the lower part of the shore, release of the capsules does not have to be synchronized with high water of spring tides. The viviparous winkle breeds throughout most of the year. The eggs are retained in the brood pouch, until the young winkles emerge and crawl into rock crevices for safety. The flat winkle lays an egg mass on brown seaweeds, including bladder and serrated wrack. The larvae develop inside the egg and the young winkles emerge directly on to the seaweed.

Life histories have evolved into a variety of strategies by which species endeavour to survive in the dynamic seashore environment.

PROJECTS

① Make a 5-cm-square wire frame. Count the number of barnacles within this square at different shore levels. At what level is the greatest concentration found? Are there any other animals encrusting the rocks at this level?

② Compare the abundance of different age groups of barnacles within the frame. Each age group will be a distinct size, ranging from recently settled barnacles to the progressively larger 1, 2 and 3-year-olds.

Using a 5 cm² frame for counting barnacles.

③ In the spring, examine rock crevices and rocky overhangs for eggs, but do NOT collect any of them. Do there appear to be any preferences for a microhabitat, or for a tidal level, by each species?

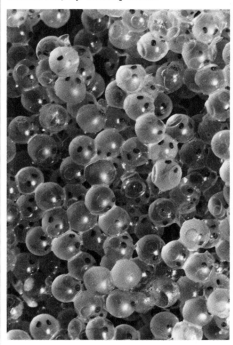

Close-up of lumpsucker *Cyclopterus lumpus* eggs, showing the pair of black eyes of each embryo. In Scandinavia, these eggs are collected and sold as mock caviar.

145

UNDER ROCKS AND IN HOLDFASTS

The undersides of boulders too heavy to be moved by waves often carry a rich fauna of both encrusting organisms and free-living ones which use it as a temporary refuge.

The space between boulders and rocks and the shore itself is an area well protected from wave action and desiccation. There is a very clear-cut distinction between the life which grows on boulder tops and that which grows on boulder bottoms. Since the upper surfaces are well exposed to light and air, only organisms which can tolerate exposure to wind and sun will be able to survive here. Seaweeds which thrive on bare rock faces will also grow on boulder tops, as will hard-shelled animals such as barnacles, mussels and limpets.

There is a sharp demarcation line between the species – particularly seaweeds – which seek light, and those which do not require light but which seek shelter and damp surroundings. Provided that boulders are not embedded in fine silt, their undersides will be covered with a rich fauna of both encrusting and free-living animals. All too often over-zealous crabbers turn over huge boulders low down on the shore in search of edible crabs and fail to return them to their original positions. This not only kills off the delicate encrusting animals by exposure to the air, but also kills the algae by grinding them into the shore.

Suspension feeders

Sessile sponges, tube worms, sea squirts and sea mats may encrust large boulders to such an extent that they form a rich, multi-coloured carpet. Boulder sponges may be yellow or green (the breadcrumb sponge), bright orange, or even navy blue. Adding to the colourful array are often small patches of colonial sea squirts, as well as the occasional specimen of the solitary sea squirt *Ciona intestinalis*. Off-white clusters of purse sponges also occur spasmodically. All these encrusting animals feed on food particles suspended in sea water.

▲Cowries *Trivia monacha* feeding on compound sea squirt *Didemnum gelatinosum* on the underside of a boulder. Both of the cowrie shells are covered by the blotched mantle folds.

▶A sea cucumber *Cucumaria normani* attached to a boulder, photographed through water. The rows of suckered tube feet run the length of the body, while the finely branched feeding tentacles project from the front end.

▲The underside of a rocky overhang encrusted with the breadcrumb sponge *Halichondria panicea*, showing the miniature volcanoes where the waste water leaves the sponge.

▼Snakelocks sea anemones *Anemonia sulcata* clinging on to a boulder which has been turned on its side. They are unable to contract their tentacles when exposed to the air.

▲A blue-rayed limpet *Patina pellucida* has eaten away so much of the upper holdfast of an oarweed *Laminaria digitata* that the whole stipe would be broken off in rough sea.

However, by no means all suspension feeders which live beneath boulders are sessile or colonial. The flattened bodies of the porcelain crabs as well as the sharp spines on the ends of their legs are both adaptations for clinging on to boulders; even the very wide claws of the broad-clawed porcelain crab are completely flattened. These crabs use their hairy mouthparts to filter off the food particles in a similar way to the attractive squat lobster, which also hides under boulders. The streamlined body of the porcelain crab offers the least possible resistance to moving water, which ensures that it is unlikely to get washed away from its boulder in rough seas. Although the plan view of a porcelain crab is not much like a mayfly nymph from a fast-flowing stream, both animals have to face the problem of maintaining their position in a fast flow, and both have a very low profile when seen from the side.

When porcelain crabs are exposed to light and air by upturning a boulder, they either drop off into the water or crawl rapidly down the boulder towards the water.

Deposit feeders
Food suspended in sea water is not the only particulate food utilized by boulder fauna. Sea cucumbers and many worms use tentacles to pick up detritus which settles on the beach surface.

Carnivores
Once the sessile growths and sluggish deposit feeders have established themselves, various carnivorous animals which feed on them are attracted to move in beneath boulders. The two cowries *Trivia arctica* and *Trivia monacha* feed on compound sea squirts, and also lay their vase-shaped egg capsules in holes within the sea squirt test (the hardened outer covering). Opportunist sea anemones outstretch their tentacles to catch any passing prey. Many sea slugs are predators which prey on a single or a few related species; for example, the sea lemon feeds almost exclusively on the breadcrumb sponge. As the sea lemon crawls under water, the delicate tripinnate gills which encircle the anus stand up from the body while, at the opposite end, the paired tentacles are outstretched. The brownish *Jorunna tormentosa* also feeds on the breadcrumb sponge; while the red *Rostanga rubra* blends in so well with the red sponges on which it feeds, that it is extremely difficult to spot. Long-legged sea spiders feed on sea anemones by secreting enzymes which break down the anemone body which can then be sucked up by the proboscis. Hydroids and small sedentary polychaetes also fall prey to sea spiders.

Unlike the filter-feeding and deposit-feeding segmented worms, the highly extensible bootlace worm is unsegmented. This ribbon worm can be found as an untidy coil beneath stones resting on muddy shingle, but when a worm is carefully straightened out, without overstretching, it reaches more than 30 metres in length. To catch its prey, the bootlace worm shoots out a long proboscis.

Predatory crabs, such as the shore crab and the velvet swimming crab, may take temporary shelter beneath boulders when exposed by the ebbing tide. Where there is a depression beneath the boulder, and the substrate is soft enough, edible crabs dig themselves into the beach surface, leaving only their eyes and antennae protruding.

Other mobile animals which can be found beneath boulders include starfish, sea urchins, brittle stars and various shore fish such as the blenny, the butterfish, the rock goby, rocklings and even the conger eel.

On gently sloping shores with a big tidal range, where sea water is retained in rocky gullies, the undersides of boulders are especially rich sites.

Holdfast fauna
The large basal attachment, known as the holdfast, of the impressive brown *Laminaria* seaweeds makes an ideal miniature hiding place for a host of small animals, and a complete oarweed forest offers countless hideaways. Compared with other seaweeds higher up the shore, a much greater variety of species lives associated with oarweeds. The diversity of holdfast fauna relates to many factors, but particularly to the extent of wave action and the amount of sediment carried in the sea water.

Worms are by far the most abundant group of animals which inhabit holdfasts. On a Devon shore in south-west England, 2,056 animals out of a total of 2,864 found in 100g of damp seaweed were segmented worms. Tiny molluscs and tube worms spend all their lives within the confines of a holdfast, while brittle stars and active, errant worms use it as a temporary shelter when the upper laminarian fringe is exposed during low water of spring tides. Encrusting sponges and compound sea squirts both grow on holdfasts, while juvenile edible crabs use it as a temporary home. Although the hairy crab is not confined to living in holdfasts, it often inhabits them. One form of the blue-rayed limpet which is confined to holdfast spaces has a rougher and much higher shell than the form which lives out on the laminarian fronds. Once in the holdfast, the limpets may excavate a miniature cave in the top of the holdfast beneath the base of the stalk.

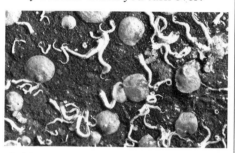

Saddle oysters *Anomia ephippium* and keelworms *Pomatoceros triqueter* are two sessile organisms which will live beneath stones low down on the shore.

The star sea squirt *Botryllus schlosseri* also encrusts rocks, but it has no hard shell and each 'star' is made of many individuals living together to form a colony.

The flattened body of the porcelain crab *Porcellana platycheles* is well adapted for living beneath boulders in currents.

LIVING TOGETHER

Several species of marine invertebrates live in close association with other species. This may be a loose association, such as hitching a ride or sharing a burrow, or a much more intimate one.

Generally, when two species live harmoniously together without either one preying on the other, the association is known as 'commensalism'. 'Symbiosis' is a much more intimate association, such as, for example, that of algal cells living inside animal tissues. 'Parasitism' is a unilateral association whereby one species preys on, or within, its host. By no means all associations are so clear-cut that they fall neatly into any of these three categories.

Plant/animal partnerships

Most associations are made between two species of animals, but there are some examples of animals and plants which live together. The growth of algae on limpet and other mollusc shells is a response to grazing pressure by the molluscs themselves, for limpets are unable to browse their own shells! There appears, however, to be no benefit to the molluscs. Although seaweeds will grow by chance on the backs of spider crabs, they are also actively planted there by the crabs to aid their disguise (pp. 142–3). Abnormal growths on fucoid seaweeds are induced by nematode worms living in the tissues. Warty galls are not uncommon on the knotted wrack.

Commensalism

Several partnerships exist between crabs and sea anemones. The spider crab *Inachus rostrata* is always found close beside the snakelocks anemone. When disturbed, the crab shelters among the anemone's tentacles, which are also used as a refuge by the Mediterranean goby. The fish is protected from the anemone's stinging cells by a mucus secretion. Larger specimens of the hermit crab *Pagurus bernhardus* which live in the sublittoral may carry sea anemones on their shell homes. These anemones are able to recognize shells inhabited by hermit crabs and they will actively move towards them and attach themselves to the shell home. The crab helps this transfer by keeping quite still. As the

◀Hermit crab *Pagurus bernhardus* may share its whelk shell home with sea anemones *Calliactis parasitica*.

▶Parasitic copepods *Pseudocaligus brevipedes* attached to outside of a 3-bearded rockling *Gaidropsarus vulgaris*. The elongated white structures are egg sacs.

▶Warty gall on knotted wrack *Ascophyllum nodosum* is caused by a parasitic nematode *Halenchus fucicola*.

▼Mussel *Mytilus edulis* opened to show pea crab *Pinnotheres pisum* inside. There is usually only a single crab inside a mussel.

Commensal worm *Nereis fucata* emerging from inside shell inhabited by a hermit crab *Pagurus bernhardus*.

crab moves over the bottom it disturbs small worms and drops scraps of food on which the anemones can feed. In return, stinging cells within the anemone tentacles help to protect the crab from predators. In the Mediterranean, another species of hermit crab *Pagurus arrosor* searches for *Calliactis* anemones which it places on its shell as a means of protection from the predatory octopus.

Other animals which may associate with the hermit crab *Pagurus bernhardus* include the colonial hydroid *Hydractinia echinata* which appears as a whitish fluffy growth on the shell; and the worm *Nereis fucata* inside the shell. By providing a hermit crab with a glass shell, the behaviour of the worm could be studied using a dim red light. The worm spends most of the time inside the upper shell coils, occasionally

emerging head-first to seize food from the crab's pincers. In return, the crab may benefit by the worm's creating a current of water through the shell. The proportion of shells which carry the worm appear to increase with a rise in temperature of the sea water. For example, in Sweden, worms were found in only 6.6 per cent of shells inhabited by crabs, whereas the percentage increases almost four times to 23.6 per cent on the west coast of Scotland and to as much as 50 per cent off Portugal.

The cloak anemone forms a much closer association with the hermit crab *Pagurus prideauxi*, which appears to be essential for the survival of each species since they never occur alone. The anemone attaches itself on to the tiny shell home below the crab's mouth, gradually surrounding the crab as its foot enlarges.

Symbiosis
Several invertebrates, both marine and freshwater, have symbiotic algal cells living inside their tissues. The attractive snakelocks anemone seeks sunlit pool margins and crevices so that the microscopic zooxanthellae living inside its tissues can utilize the sun to build up carbohydrates and release oxygen by means of photosynthesis. The plant cells gain protection within the anemone, which itself benefits by an increase in the available oxygen in pools exposed by the ebbing tide.

The brilliant green marine flatworm *Convoluta roscoffensis* (pp. 122–3) owes its colour solely to symbiotic dinoflagellate algae. The algae do not occur in the egg, but the young, colourless worms are soon infested. The digestive organs of the worms degenerate as they begin to feed on the starch produced by the green algae. Eventually, the flatworms kill the hand which feeds them when they become parasitic, feeding directly on the algal cells. Shortly afterwards, the flatworms lay a large number of eggs, before they die.

Parasitism
Animals which feed on the outside of their host are known as ectoparasites; organisms which live inside their host are endoparasites. Parasitism is very much a one-sided association, with the parasite gaining and the host losing to a lesser or greater extent. Unlike predators, parasites rarely kill the animal on which they feed; for if they did they, too, would die. While ectoparasites tend to show some modifications such as hooks or suckers, for clinging on to their host, their bodies are much less modified for a parasitic existence than are those of endoparasites.

On the shore, parasites most likely to be encountered are external copepods on fish, and the internal sac barnacle which infests shore crabs. The most conspicuous parts of caligid copepods, which are known as fish lice, are the large pair of egg sacs which project far beyond the end of the body.

Infestation of a crab by the sac barnacle is apparent only when the parasite produces its fleshy yellowish or white egg sac beneath the crab's abdominal flap. At first glance, it could be mistaken for a female crab's own egg mass, but the sac barnacle attacks both sexes, and its egg mass is not granular. Long before the external egg mass forms, the parasite grows inside throughout the crab as a branching network. The parasite can prevent reproduction altogether by destroying the gonads and then male crabs undergo a sex change to become females. Hermit crabs are attacked by another parasitic barnacle, *Peltogaster paguri*, which can be detected only when a hermit crab is removed from its shell home.

Also found on rocky shores—inside mussels—is the tiny pea crab, which leads a semi-parasitic life by stealing planktonic food from the mussel's gills. Pea crabs are well protected inside the mussel, and so they have only thin shells, and feeble legs, but they have elaborate fringes on their claws for combing the food trapped in the mucus. The mussel is weakened by the loss of food and it may also have its gills damaged. Female pea crabs, which are bigger than the males, may grow so large that they become imprisoned inside the mussel. This is no great hardship, since the small male can still visit her and mate. The behaviour of a pea crab inside a mussel has been observed through a window cut in the shell.

PROJECTS

1 Collect shells inhabited by hermit crabs which have white fluffy growths on them. Transfer to a sandwich box with sea water. Examine the growths with a hand lens, so that the individual polyps of the hydroids can be seen.

Sharing a hermit crab shell is the delicate colonial hydroid *Hydractinia echinata*. When placed under water, its stalked polyps are clearly visible.

2 Examine shore crabs to see what proportion are parasitized by the sac barnacle. Such crabs are easy to spot because the parasite prevents them from moulting and so their shells will be encrusted with barnacles and other growths.

Egg mass of sac barnacle *Sacculina carcini* visible beneath abdominal flap of shore crab *Carcinus maenas*.

3 How many different species are hitching a ride on the backs of the crabs?

This shore crab *Carcinus maenas* is infested with the parasitic sac barnacle *Sacculina carcini* which has inhibited moulting, as can be seen by the mass of acorn barnacles on the back of the crab's carapace.

ROCK POOLS

Pools on rocky shores are
attractive natural aquaria where a
selection of seaweeds and shore
animals can be observed during
low tide. However, in reality,
these habitats are very
harsh ones.

The origin of pools

The reason that some rocky shores have
a great many intertidal pools, while
others have few or none at all, relates to
the geology of the shore. The alignment
of the rock strata and the way in which
the rock is eroded and weathered are
factors which affect the formation of
pools. Their shapes and sizes vary
greatly – from narrow slits between
uplifted strata to shallow, sunny pools
on wave-cut platforms and deep,
shaded pools low down the shore. Pools
form wherever sea water remains
trapped after the tide recedes. Softer
rocks, such as limestone, which erode
more readily than hard granite, harbour
copious pools of all sizes.

The pool environment

Although life in pools does not have to
face the problem of desiccation, it is
none the less subjected to much wider
extremes of temperature and salinity
than the open sea. High-level pools,
which are exposed to the air for long
periods – as much as several consecutive
days – in between submergence,
experience daily temperature, oxygen
and pH fluctuations.

High-level pools

On warm, sunny days the temperature
of the water in shallow high-level pools
can rise to more than 10°C above the
sea-water temperature, only to be
suddenly lowered again when the
incoming tide flushes out the pool. On
the French Atlantic coast, however, two
Mediterranean fish – the giant goby and
Montagu's blenny – flourish in the
warm water of high tidal pools, but
cannot survive lower down the shore.

Pools with a luxuriant growth of the
green alga *Enteromorpha* are 1–2°C
warmer than pools at the same shore
level with little or no algae. These
plants also give rise to a high oxygen
level (as a by-product of photosynthesis
in the presence of sunlight) and a high
pH level by day; and a high carbon
dioxide level by night, when the algae
no longer photosynthesize but continue
to respire. On a cold winter's day, when
the air temperature is below the sea

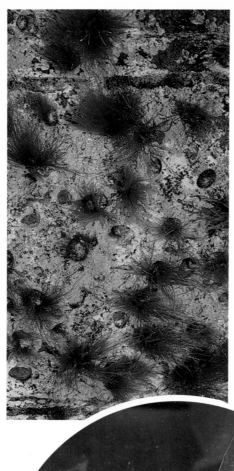

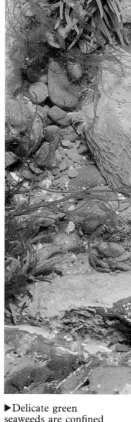

◄A detailed close-up
of a pool bottom
reveals individual pits
in the soft limestone
made by limpets. The
green algae are
confined to the limpet
shells where they
cannot be grazed.

►A depression in red
sandstone rock, on the
lower part of a shore,
harbours a rich
assortment of marine
invertebrates,
including mussels,
limpets, winkles, a
juvenile shore crab,
snakelocks sea
anemones and a sea
slug *Aeolidia papillosa*.
The bottom of the pool
is covered with
encrusting algae.

▼A chameleon prawn
Hippolyte varians is
perfectly camouflaged
against green sea
lettuce *Ulva lactuca*.

►Delicate green
seaweeds are confined
to a pool in the middle
of a rocky shore, while
the brown wracks
tolerate several hours
of exposure on bare
rocks, where they turn
almost black as they
dry out on a hot
summer's day.

▼The highly eroded
limestone shore in the
Burren, on the west
coast of Ireland, is
covered with pools of
all sizes. In the bottom
of this one, rock
urchins *Paracentrotus
lividus* have made their
own pits by rotating
their spines against the
soft rock.

temperature, the incoming tide will warm the pool water.

When pools containing a similar volume of sea water at different shore levels are compared, high-level pools experience much greater salinity fluctuations. When evaporation takes place on a hot day the salinity increases, whereas if rain falls during a low tide it will reduce the salinity.

The range of animals living in high-level pools is therefore poor, but those species which have adapted to temperature and salinity fluctuations are present in large numbers. *Heterocypris salinus* is a small ostracod (a bivalve crustacean) which is negatively phototactic and so it moves away from the light down to the pool bottom by day and up to the surface waters at night. The copepod *Trigriopus fulvus* shows a remarkable salinity tolerance, being able to live normally at salinities ranging from 4 to 90 parts per thousand.

On the surface of high-level pools huge swarms of tiny bluish insects occur. These 2mm-long marine springtails, which have hairy, non-wettable bodies, are the only shore insects living on the water surface, whereas in freshwater habitats several species have exploited the surface film. When the incoming tide submerges the pools, the springtails crawl into crevices where they survive in tiny air pockets.

Mid-shore pools

Some of the most attractive pools are the shallow ones lined with the pink encrusting algae *Lithophyllum*, and interspersed with scattered tufts of the seaweed *Corallina*. Limpets live both in and out of pools; when those from coralline pools die, they leave behind distinctive circular areas of bare rock marking their 'homing' positions where algal growth has been prevented. Unlike beadlet sea anemones exposed on bare rock, anemones in pools can keep their tentacles expanded. Snakelocks anemones are unable to contract their tentacles and so are confined to pools and gullies.

Many crustaceans live in pools, the most obvious ones being the hermit crabs which inhabit discarded winkle shells. Even if the crab is not fully extended from its shell home, it can be detected by the speed at which it moves, because this is much faster than a live winkle.

Low-level pools

Pools low down the shore are exposed for only short periods and therefore show less dramatic temperature fluctuations than high-level pools. Large, deep pools are places in which sublittoral species can extend their range higher up the shore and still survive between the tides. Prawns are so well camouflaged that they are not easily seen by day unless the seaweeds are gently moved to dislodge them. When alive, edible prawns are transparent, and so blend in well with the weedy fringes; but the chameleon prawn is able to change its colour by day to blend in with the seaweeds.

As well as the regular inhabitants of pools, fish, crabs and the occasional starfish or sea urchin may get stranded in a pool as the tide recedes.

Pool avoiders

There are several open rock organisms which are completely absent from pools. This is best shown by the acorn barnacles, which end as an abrupt line immediately above the pool surface. The reason for this absence of submerged barnacles may be due either to the failure of their larvae to settle in pools, or to the adults' inability to tolerate continuous submergence. Small winkles and thick topshells are rarely submerged in pools.

PROJECTS

1 Watch life in a pool, either wearing a diver's face mask or using a small perspex aquarium. An underwater viewer can be made by removing the bottom of a large plastic tub or a tin (make sure there are no sharp edges) and covering the bottom with a clear sheet of plastic held in place with a rubber band. Lie on the rock beside the pool. Put the mask, the aquarium or the viewer just below the surface and look down into the pool. If you keep quite still, you will see small fish and prawns darting out from their seaweed cover.

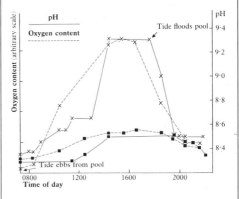

2 Visit a sheltered, rocky shore on a warm, sunny day with a thermometer and a book of pH papers. Take the temperature and the pH of the water in pools at different shore levels. Sketch the shore so that the pools can be found again at a later date.

× Pool containing abundant algae
■ Pool containing few algae, above MHWS

Comparison of oxygen and pH content of two tidal pools, one of which had abundant plant growth (×) and the other only sparse growth (■).

3 Return to the same area on a cold, overcast day. Take the temperature and the pH of the same pools. Compare the pairs of readings for each pool by plotting them on graph paper. Explain any differences between the pairs of readings.

CLIFFS

Cliffs, arches and stacks are spectacular features of coastal scenery which all arise from the erosion of land. Arches arise when the sea breaks through a narrow headland; stacks are tall columns of rock which have been separated from cliffs. The profile of a cliff depends on the rock type and how exposed it is to the elements. Where a coastline is irregular, the force of the waves will tend to be concentrated on headlands which, if made of hard rocks, will resist erosion better than soft rocks backing neighbouring bays. As a cliff face is eroded away, so the profile of the coastline changes.

Formed by the erosive powers of waves, wind, frost and rain wherever high land and sea meet, cliff faces and stacks are usually inaccessible to man but not to birds. These sites are often alive with nesting sea birds such as guillemots, razorbills, kittiwakes, gannets and fulmars. Ledges provide valuable nest sites, as well as places for cliff plants to gain a foothold. As cliffs erode away, rocks fall to the base where they are ground down by the sea. It is these rock fragments which build up to form sand and shingle beaches. As gales are such a regular feature of coasts—especially cliff tops—no trees grow close to cliff edges, and even inland trees are wind-pruned into lopsided shapes.

The Green Bridge of Wales is a natural arch made of carboniferous limestone which was laid down on an old sea floor when the sea level was 60 metres higher than it is today.

CLIFFS, ARCHES AND STACKS

Cliffs develop from the erosion of land by sea, rain, wind and frost. Marine cliffs include not only hard, rocky walls, but also soft, unstable sand and clay profiles, as well as chalk faces which tend to crumble away easily.

Origin of rocks

Rocks are formed in quite different ways. Igneous rocks, which arise when molten magma and volcanic lava cool, include crystalline basalt, granite, mica and quartzite. Sedimentary rocks form by the accumulation either of weathered particles swept down into shallow seas (sandstones, grits and conglomerates); or of the remains of shells, corals and skeletons (limestones and chalk); or as a result of chemical precipitation (travertine and oolite limestone). Metamorphic rocks form as a result of high temperatures and pressures changing the structure of existing rocks.

Once formed, the nature of rocks may be further changed by movements of the earth's crust resulting in folding and faulting. Faults arise when cracks develop in rocks which are then pushed upwards or sideways.

Erosion of cliffs

The composition of coastal rocks determines not only the colour of cliffs, but also their rate of erosion. Softer and weaker rocks are most vulnerable to attack. The rate at which cliffs are worn away also depends on the topography of the land, the orientation of the rock strata in relation to the sea and the amount of rainfall as well as the degree of exposure to the sea. Places where soft cliffs are eroding away quite rapidly can be seen along the coast in south and east Britain, in north France, in Denmark, and along the low Baltic coast.

In stormy seas, huge blocks of cliff face may fall away as joints (vertical cracks) are weakened and attacked by powerful waves. If the fallen rocks are then hurled against the base of the cliff, they will undercut it. Slips are a feature of the coastline between Dover and Folkestone in Kent in south-east England and along the south coast of the Isle of Wight, where the chalk lies above the gault clay. As the clay becomes slippery it can no longer support the rocks above it and so a cliff

The white chalk cliffs of the Seven Sisters in Sussex, in the south of England, contrast with the chalk cliff in the foreground which is coloured with a capping of lower Tertiary sands. On the beach is ample proof of recent cliff falls.

▶Outline drawing of the colour plate of the Seven Sisters, showing geological features.

CHALK CLIFF PROFILE
HANGING VALLEYS FORMED AS VERTICAL CLIFFS CUT BACK
TERTIARY SAND CAPPING ON CHALK CLIFF
WAVE CUT PLATFORM
BOULDERS
CHALK AND FLINT NODULES
UNDERCUT CLIFFS WHEN THROWN AGAINST IT BY WAVES

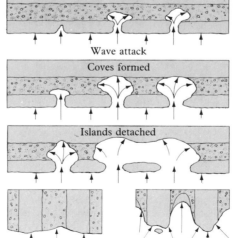

▲The walls of this cave in south-east Iceland are formed from volcanic basalt lava columns, which provide numerous ledges for cliff plants to grow.

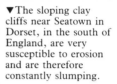

▼The sloping clay cliffs near Seatown in Dorset, in the south of England, are very susceptible to erosion and are therefore constantly slumping. As the cliffs crumble away, fossil ammonites get washed down on to the beach. At the base of the cliffs, the sea has thrown up a shingle ridge.

Weaknesses exploited

Wave attack

Coves formed

Islands detached

Bay with bay-head beach   Headland

▢ Resistant rock   ▢ Less resistant

▲Diagrammatic representation showing how the orientation of the rock strata, in relation to the shore line, affects the erosion, and hence the coastal topography.

fault results. An example of a very spectacular coastal landslip can be seen at Axmouth – Lyme Regis Undercliffs in Devon in south-west England. On Christmas day in 1839, nearly eight million tons of chalk and upper greensand rock split away from the cliff, slipping towards the sea on the underlying gault, producing a chasm 30–61 metres deep and 61–122 metres wide. The slipped land is now covered with a natural ash woodland.

Large cliff falls and landslips tend to occur most often during the winter months when the rainfall is high and frosts as well as gales occur. Concrete sea walls and promenades are built to protect vulnerable cliffs, especially in seaside resorts, from more rapid attack by the sea.

The rate of cliff-top retreat can be seen by comparing the mean annual loss in different sites in the south of England. The Tertiary beds in north Kent are eroding at a rate of 0.81 metres per year; cliffs in Brightstone Bay in the

▲The top and every available ledge of a tall columnar stack off one of the Farne Islands, off the north-east coast of England, is utilized by breeding sea birds.

▼Archways are a feature of the Algarve coastline near Portimao in Portugal which has been shaped by the erosive powers of the sea.

Isle of Wight at 0.51 metres per year; and the Sussex cliffs at 0.38 metres per year. Boulder clay cliffs are soft and therefore easily eroded, whereas hard rocky cliffs made of basalt, granite, limestone, sandstone, serpentine or slate are comparatively stable and more resistant to erosion. They are, even so, being constantly slowly eaten away. Coastal rocks are shattered by pounding seas, especially if rock fragments are carried by the waves or if the air in rock holes and crevices is suddenly compressed by wave action. Once the sea weakens the rocks, they are further eroded by wind and rain and by frost shattering. High cliffs are formed after a long period of erosion has cut back gently sloping bedrock on the seaward side.

In Britain and Ireland, cliffs are very much a feature of the older, harder rocks found along the south-west coasts, and also in the north of Scotland. Even more spectacular cliffs occur in north-west Iceland. Exposed to the full battering of the Norwegian Sea, the Tertiary basalt forms vertical cliffs 500 metres high, which are colonized from top to bottom by millions of breeding seabirds, including many thousands of Brünnich's guillemots. It is the guano from all these birds which supports a rich plant community on the cliff ledges.

Headlands

When hard, resistant rocks alternate with softer rocks at right angles to the shore, the soft ones erode away much faster, leaving the hard rocks jutting out as headlands. More rapid erosion of the soft rocks gives rise to bays with beaches. One of the most bizarre rocky headlands must be the Giant's Causeway in County Antrim, Northern Ireland. Here, the spectacular hexagonal basalt columns, which are of volcanic origin, form one of the world's geological scenic marvels.

Arches and stacks

Arches and stacks develop as the sea enlarges weak joints or faults in rocky headlands. An arch forms when the sea attacks both sides of a narrow headland to form caves, which eventually meet, leaving the overhead arch or bridge. A famous sea arch is Durdle Door in Dorset. Other places where arches can be seen are near Portimao, on the Algarve coast in Portugal, at Land's End in Cornwall, and on the Tenby Peninsula in south Wales.

When an arch collapses, a column or sea stack is left standing as a small columnar island adjacent to the coast. Sometimes, a stack may become connected to the mainland during low tides when the sea retreats to expose a wave-cut platform. The highest sea stack in Britain is the 191-metre-high Stac an Armin in the St Kilda group. Spectacular old red sandstone stacks rise 90 metres from the sea off Duncansby Head at the north-east tip of Britain; while the Old Man of Hoy in Orkney reaches 137 metres high. The chalk Needles off the Isle of Wight may not match these northern stacks for size, but they are a familiar landmark to sailors the world over.

Caves and geos

As joints enlarge, caves form in rocky cliffs, and repeated compression of water against the cliff walls and ceiling can result in the collapse of a cliff roof to form a blow-hole. When the ceiling of a cave collapses along its entire length, an upright narrow chasm—a geo— forms. Geos are a feature of the Shetland coastline and the west coast of the Orkney mainland. It is a spectacular sight when storm waves crash into these narrow rocky inlets, especially in the Irish geo known as Moista Sound, 122 metres high and 186 metres deep.

PROJECTS

① Make an outline sketch of a cliff scene, identifying all the features which you can see, as in the sketch at the top of the page opposite. Relate these features to the type of rock and the exposure to wave action.

② Compare a detailed modern map of a stretch of coastline with soft cliffs, with an old map. What changes have taken place? Also, look at old paintings or postcards of well-known coastal scenery. Try to find precisely the same viewpoint as on the old illustration, so that you can compare the two, and, if possible, take a photograph. Are the differences between the old and the present-day view subtle or obvious? Try to explain the reason for this.

Above: Pulpit Rock, Portland, in the south of England, from a postcard taken in 1948.
Below: Pulpit Rock photographed 33 years later in 1981. Can you spot the differences between the two photos?

SEABIRDS

Seabirds prefer to breed on cliffs and stacks which are not disturbed too much by man. Some of the largest concentrations therefore occur on remote offshore islands.

Although seabirds feed at sea, they must come ashore to lay their eggs and rear their offspring. Cliffs, stacks and islands along the Atlantic and North Sea coasts from as far north as Greenland and Iceland, down to Britain, are utilized by breeding seabirds, which generally nest in colonies. Some of these colonies may extend to several thousand pairs.

The large seabird colonies in north-west Europe coincide with the warm and productive North Atlantic Drift current. The cliffs of the British Isles, north-west France and Norway support more than three-quarters of the world's gannets and razorbills. Most seabirds do not remain on their breeding grounds throughout the year.

Once young seabirds are reared, the adults leave the nests during the winter months, returning again in early spring. In March, large flocks of auks (puffins, guillemots and razorbills) assemble in rafts on the sea surface beneath the cliffs.

Nest sites

The noise and smell of a large seabird colony, as well as the constant movement of the adults leaving the nest and returning with food, is quite unforgettable. Seabirds avoid competition when nesting by showing distinct preferences for certain niches.

Shags and cormorants build their nests near to the sea on ledges and flat-topped stacks up to 20 metres high, which makes it easier for them to return with sodden wings.

Higher up, kittiwakes and guillemots breed on narrow ledges. Kittiwakes build a large, cup-shaped nest from plant debris and mud which is attached to the ledge by a cement made from green alga and guano. Guillemots are highly gregarious birds, packing themselves in tight rows on ledges or in large groups on top of stacks. They are then less likely to take fright and thereby lose their eggs laid on bare rock by suddenly lifting-off.

Fulmars prefer broader ledges for nesting at a density of 1–2.4 nests per square metre. If young fulmar chicks are disturbed, they will regurgitate an

▲Contrasting plumage of the adult (white) gannet *Sula bassana* and the chick (black).

▼Different techniques used by seabirds to obtain their food.

Surface plunging

Gannet

Dipping

Gull

Pattering

Storm-petrel

Pursuit diving with wings

Cormorant

Auk

Pursuit diving with feet

▲A herring gull *Larus argentatus* trumpeting. This elaborate challenging call can be heard throughout the year, but is most frequent and intense during the reproductive season.

unpleasant-smelling, oily liquid. Gannets also prefer broad, flat cliff ledges as well as the more extensive cliff tops, nesting close together in large gannetries. The volcano-shaped nests are made from seaweed, feathers and flotsam, cemented together by excreta. Within the last thirty years new colonies have been established in Iceland and gannets now breed in both Norway and France.

Gannets, like most seabirds, mate for life. Even after two birds pair, they continue to perform displays to each other, which strengthens the pair bond and drives away intruders. When a bird returns to its nest after a feeding foray, the two will greet each other by 'sky-pointing', the tips of the bills being touched on alternate sides as they are

▲A fulmar *Fulmarus glacialis* nesting on a rocky ledge in the Farne Islands with sea campion flowering below.

▼Kittiwakes *Rissa tridactyla* nesting on narrow, rocky ledges on a Farne Island stack.

pointed skywards. When it comes to defending their territory, gannets are ferocious fighters. The signal they use to assert ownership of their nest site is an elaborate bow.

Gulls are much more adaptable in their breeding sites than other seabirds, and in recent years herring gulls have begun to invade high-rise buildings in towns and cities.

Puffins and Manx shearwaters are both colonial nesters which breed in burrows, usually on cliff tops or slopes. Old rabbit burrows are often used by both these birds. In a dense colony, where the ground is riddled with burrows, the surface is very friable, and so great care must be taken not to cause the burrows to collapse by walking on top of them. Manx shearwaters will not

be seen by day, as they are completely nocturnal.

Feeding

Seabirds feed on a wide variety of food from plankton to crustaceans and fish, but each species prefers a certain food type. The method used for catching food varies greatly. Guillemots, razorbills and puffins feed on small fish such as sprats and sand-eels by diving from the surface and flapping their wings as though 'flying' under water. During the breeding season, puffins can be seen returning to their burrows with a neat beakful of sand-eels. Inside the brightly coloured mandibles are backward-pointing serrations which help to stop the fish falling out.

Cormorants and shags use their feet to propel themselves through water while feeding.

Gannets make spectacular plunge dives, head first into the sea from a height of over 30 metres. They have extra strong skulls to help absorb the shock when they hit the water. The fish is grasped by the bill and swallowed whole as the gannet breaks the surface. Kittiwakes will plunge-dive for fish, as well as feeding at the surface when sitting on the sea or flying low over the water.

Storm-petrels feed on zooplankton and small fish by hovering above the sea, pattering their feet along the surface and picking up food in their bills. Fulmars feed mainly on zooplankton, but they are also efficient scavengers.

Seabirds as food

Young seabirds have a much higher mortality than adults, which are long-lived. For example, a gannet can live for over 23 years, and a fulmar for over 40. The gannet has a mortality of about 70 per cent in the first year of life. As well as starvation, severe weather and accidents, many eggs and young seabirds fall prey to predatory birds, such as herring gulls, lesser black-backed gulls and Arctic skuas.

On remote islands, in particular, seabirds and their eggs were an important source of food for man. Seabird eggs are very nourishing; weight for weight, guillemot eggs contain more calories than beef. The entire culture of the self-sufficient community of St Kilda, off north-west Scotland, was based on seabirds. Working in teams, men scaled the high cliffs on horse-hair ropes to collect young gannets, puffins (for feathers) and fulmars which provided oil for lamps. As many as 5,000 gannets, 20,000 puffins and 9,000 fulmars could be taken in a year. After the feathers were plucked, the birds were salted and stored in small stone buildings known as cleits. Seabirds are still collected today in the Faroes.

PROJECTS

①　Find out where seabirds nest and visit the site during the breeding season with binoculars, a watch, a field guide for identifying birds and a notebook.

A sketch of a pair of guillemots *Uria aalge* greeting by bill-touching.

②　First of all, identify the different birds; then spend some time observing the behaviour of one species. A visit to a gannetry will show you all kinds of behaviour patterns. In what ways does a bird display to its mate? How frequently do these displays take place? How do birds react to neighbouring birds encroaching into their territory?

③　What food do birds bring back to the nest? How frequently do the adults return with food? Is it fed to the young fresh or regurgitated?

A puffin *Fratercula arctica* with a sand eel in its bill.

④　While you are using binoculars look to see if any of the birds' legs have been ringed. Ringing is a way of marking individuals to see if they return to the same nest site and also to see how far they migrate away from their breeding site.

CLIFF VEGETATION

Maritime lichens colonize bare vertical rock faces, whereas flowering plants will gain a foothold only on ledges or in crevices where soil pockets accumulate. When cliff plants are flowering, they provide a spectacular natural rock garden.

The growth of cliff plants is limited by the aspect of the cliffs to prevailing winds and sea spray; by grazing animals; by the geological structure of the underlying rocks; by the angle of slope; and by competition with other plants. Earth cliffs, which tend to be low-sloping and unstable, have a thick vegetation cover until there is a landslip, when the cliffs are completely bared. Steep, high, rocky cliffs, on the other hand, have a much sparser vegetation, which is confined to ledges and crevices.

Cliffs which are easily weathered support a richer flora than hard granitic rocks. The first plants to colonize bare cliffs are algae and lichens. When these pioneers die, they get broken down into humus which accumulates in crevices to form a valuable soil pocket. Where rocks such as shales, sandstones and slates have inclined bedding planes, and erode into a series of step-like ledges, they provide valuable niches for plants to grow and seabirds to nest on the lee side of headlands.

Prevailing winds

Constant buffeting of rocky cliffs by prevailing winds severely restricts the plant cover, as can be seen by comparing the plants growing on opposite sides of a headland. Winds not only dry out the remnants of soil, but also encourage water loss from the plants themselves and whip up salt spray. Maritime plants are best adapted for life in such an extreme climate, and the flowering plants which grow on cliffs nearest to the sea are perennials with long woody taproots and succulent or leathery leaves with waxy or hairy coatings. As well as thriving on rocky headlands, thrift is one of the few plants able to hang on and survive in such unstable habitats as the summit of Ben Nevis (1268 metres) or the morainic edge of ice caps. A good example of the importance of aspect to the successful colonization of cliffs can be seen by

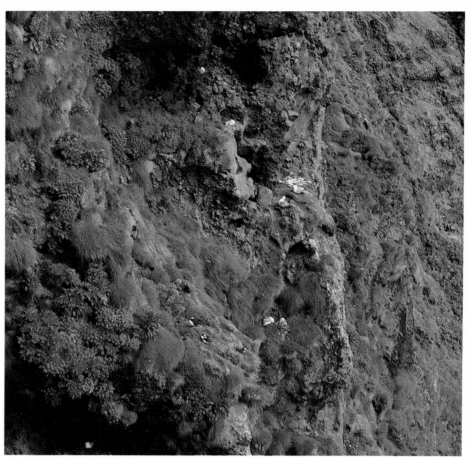

▲A luxuriant vegetation cover grows on cliffs enriched with the droppings of breeding seabirds. Here, roseroot *Rhodiola rosea* and angelica *Angelica archangelica* thrive among nesting puffins and fulmars in south-east Iceland.

◄Wild cabbage *Brassica oleracea* grows out from a crevice of a calcareous sea cliff. This plant is the ancestor of the garden cabbage.

►A clump of sea campion *Silene maritima* gains a foothold where humus has accumulated in the joints between basalt columns.

◄The bright pink flowers of thrift *Armeria maritima* give a striking splash of colour on the cliff top at Land's End, Cornwall, in south-west England.

►Common scurvy grass *Cochlearia officinalis*, which flowers early in spring, was once eaten as a source of Vitamin C.

comparing the green, south-facing White Cliffs of Dover in Kent in Britain with the white, north-facing cliffs of Cap Gris Nez in northern France. Both are built of the same chalk. The warm, sunny Dover cliffs, which are eroding at a slow rate, have a luxurious green cover; whereas the cold French cliffs are buffeted by winds and waves beating southwards down the Channel. Aspect also affects the rate of photosynthesis and hence the growth of plants.

Lower cliff plants

Moving up a rocky cliff face from the sea, the first plants to grow above the upper limit of the seaweeds will nearly always be the salt-tolerant species of lichens. These are often visible on the bare rock face as distinct coloured bands or zones: firstly, a black band (encrusting *Verrucaria maura* and tight tufts of *Lichina confinis*) followed by an orange band (*Xanthoria parietina*) topped with a grey encrusting band (*Ochrolechia parella*) and sometimes grey-green tufts of sea ivory. Intermingling with, or more often above, the lichens a limited range of flowering plants grows out from the crevices. In northern Europe thrift is the most ubiquitous cliff plant. Other crevice plants in southern Britain and Europe include sea campion, rock samphire, cliff sea-spurrey, golden samphire, buck's horn plantain, scurvy-

grass, the grass red fescue and in moist, shady crevices the fern sea spleenwort. In Scotland, west Ireland, and Iceland these plants tend to be replaced by lovage, Scottish scurvy-grass and roseroot which also grows in crevices up to 1173 metres on mountains.

Plants associated with nesting seabirds

When large numbers of seabirds nest on cliffs, they enrich the cliff soils—especially the nitrogen and phosphorus content. This tends to modify the range of plants which grow there, and to favour a few vigorous species, including scurvy-grass, sea beet and orache, which grow much larger than usual. In addition, phosphate and nitrate demanding inland plants such as stinging nettle, chickweed, goosegrass and sorrels grow in among nesting seabird colonies on cliffs. The striking tree mallow often grows where seabirds nest.

Upper cliff plants

On the lower cliff faces, the effect of salt spray is more limiting to flowering plants than the acidity or alkalinity of the bedrock. Higher up, out of reach of the salt, the number of plant species increases and includes several plants which frequent walls in coastal sites: plants such as pennywort, pellitory-of-the-wall and ivy.

On dry, non-calcareous faces, the cover often merges into a submaritime heath on the upper slopes. On the Lizard Peninsula in Cornwall, Cornish heath grows in large clumps with other heaths on the serpentine rocks. This Lusitanian species, which also grows in France and Spain, requires a mild, oceanic climate.

On basic rocks, a variety of plants flourish including sea carrot, biting stonecrop and kidney-vetch, which help to create the colourful rock garden mosaic. On the upper faces of chalk and limestone cliffs, many inland calcicole (chalk-loving) plants grow; including the localized early spider orchids among the turf on the cliff top. Although the untouched coastal strip between the cliff edge and the cultivated land can be very narrow, if it is inaccessible to man and to stock this undisturbed habitat forms a valuable refuge for wild flowers and shrubs.

Cliff tops

In the west of Britain—especially on the offshore islands—cliff tops are bedecked in spring with blue carpets of bluebells or spring squills.

On cliff tops which are exposed to Atlantic gales, few trees or shrubs can grow near to the sea; but on the lee side of a stone wall or even a boulder, quite tender plants can grow in a sheltered microhabitat.

1 Visit an exposed cliff site and a sheltered fjord or a sea-water loch. How close to high water level do flowering plants grow in each site?

In a sheltered Irish sea-water lough, thrift grows immediately above the limit of the brown seaweeds.

2 Make a sketch of each cliff profile and, using symbols to represent the different species of lichens and flowers, plot their distribution in each site. How many species occur in both sites? Do they occur in distinct zones? It may be easier to distinguish the plants by sitting on a cliff top (do not go too close to the cliff edge) and looking down with binoculars. Can you make any conclusions about the geology or the aspect of your chosen cliffs and their plant cover?

Well above the highest seaweeds, a variety of lichens grows on the bare rock face.

3 Are there any obvious differences between opposite sides of a headland which face in different directions? On accessible cliffs with a safe cliff path, use a skewer to measure the soil depth in different places adjacent to the path. Does the amount of soil affect the plant cover?

THE OPEN SEA

Gannets, flying away from their breeding ground over the sea to find their food, illustrate the importance of the interaction between the air and the sea that dominates the ecology of the open sea. Sunlight is the fundamental requirement for all plant growth and the heat which it generates is a vital energy source for oceanic processes. One of the other sources is the wind, which whips up the sea surface into waves and also drives ocean currents. The currents are influenced by the spin of the earth, and the gravitational pulls of both sun and moon create the tides which dominate the distribution of life on the sea bed as well as on the shore.

Life in and around the sea is dominated by these physical factors far more than in most terrestrial environments. In the shallow shelf seas that fringe the continents, mixing of the water by the tides helps to maintain higher concentrations of plants than in the open ocean. The bulk of these plants are microscopic floating organisms which are stimulated to grow by the greater quantities of nutrient salts such as nitrates and phosphates which are dissolved in the sea water. Consequently, 90 per cent of the world's fisheries are concentrated in a mere 5 per cent of the earth's surface that is covered with these fringing seas; whereas over 70 per cent of the earth is covered by all the oceans.

Part of the extensive gannet *Sula bassana* colony on the island of Grassholm, 19 km off the south Wales coast. These birds expertly plunge-dive for fish in the sea.

A DRIFTING LIFE

Life which drifts freely in the ocean is collectively known as 'plankton'. It is carried passively to and fro by the ocean currents, although individuals are capable of some movement, particularly vertically up and down in the sea.

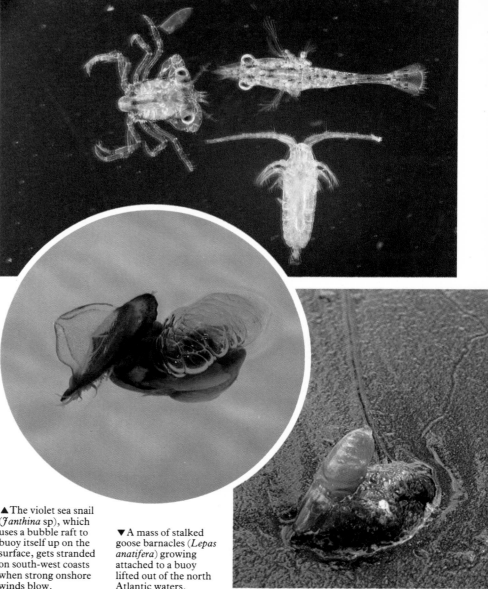

Little pinpoint flashes of blue-green light can often be seen in the surf and its backwash by visiting a beach on a warm evening in late summer. There is something quite magical about swimming or rowing in a sheltered sea-water loch at night; each dip of an arm or an oar sets off a miniature fireworks display. These flashes are produced by tiny, single-celled organisms, *Noctiluca*, which spend their lives drifting freely in the water. They belong to a group of living things that are both plant-like and animal-like. The plant-like organisms, the producers, contain chlorophyll and so are able to utilize the sun's energy in a similar way to seaweeds and flowering plants. The animal-like organisms, the consumers, eat other cells and often contain little, if any, chlorophyll.

The producers

In the open sea, where there is nothing to which large seaweeds can attach themselves, the plants have to be free-floating and are termed 'phytoplankton'. They range in size from minuscule cells, so small that they can hardly be seen even with the most powerful light microscope, to *Noctiluca*, which reaches a millimetre in diameter. Diatoms are a group of phytoplankton which have outer skeletons made of silica etched in exquisite patterns of lines and pits. In some types the skeleton is extended into long spines and processes, while in others it is formed into a pill-box structure. So regular and fine is the sculpturing on the skeleton that early microscopists used preparations of pill-box diatoms to test the optical quality of their lenses.

Winter storms bring fresh supplies of mineral nutrients, such as nitrates and phosphates, up to the surface from the depths. These nutrients, followed by a rise in the sea-water temperature and increasing day length in spring, stimulate a massive growth in the population of the diatoms.

In the summer, the diatoms tend to

▲The violet sea snail (*Janthina* sp), which uses a bubble raft to buoy itself up on the surface, gets stranded on south-west coasts when strong onshore winds blow.

▼A mass of stalked goose barnacles (*Lepas anatifera*) growing attached to a buoy lifted out of the north Atlantic waters.

die away and to be replaced by another phytoplankton group – the dinoflagellates. As with the diatoms, and also some freshwater planktonic organisms, dense concentrations, known as 'blooms', can arise, and may be so abundant that they discolour the water. Those which turn sea water red are known as 'red tides', and are so poisonous that they can cause mass mortalities of seabirds and fish. Sometimes human fatalities occur from eating shellfish which have consumed and concentrated these toxic plants. As well as discolouring the water, blooms cut down the penetration of light through water.

The consumers

All these phytoplankton are in turn grazed by animal plankton, or 'zooplankton'. These herbivorous planktonic animals range in size, from tiny ciliated protozoans scarcely bigger than the plant cells they are eating, to large relatives of sea squirts, called salps, which can grow to almost 10 cm

◄ Assorted marine plankton, including crab megalopa larva, copepod and decapod larva.

▼ Even when it is stranded, a Portuguese man-o'-war (*Physalia physalis*) can still sting a man.

▲ Sargassum weed drifts around on the surface of north Atlantic waters buoyed up by its air bladders. Among clumps of this weed lives a community of animals which blend in perfectly with both the colour and the shape of the seaweed.

long. Some of these herbivores are permanent members of the plankton, but others are the larval stages of shore- or bottom-living animals. The latter live as plankton for a few weeks or only a matter of days before they settle and metamorphose into their adult forms. These larvae are important guides to the evolutionary relationships within different groups. For example, the similarity between the early larvae of molluscs and annelid worms suggests that they share a common ancestry. Hard-shelled barnacles were for a long time thought to be a kind of mollusc, until their larvae were discovered, which showed them quite clearly to be crustaceans, related to crabs and copepods.

Copepods are the most abundant members of the permanent plankton, and many of these small planktonic crustaceans are herbivores. They comb out the phytoplankton with their legs, which are armed with a delicate series of fine bristles. Other copepods are either detritus feeders, picking up food scraps or excreta, or carnivores. The abundance of one copepod in particular, *Calanus finmarchicus*, determines how well many larval fish – notably young herring – survive.

Herbivorous crustaceans comb the water for the phytoplankton, whereas salps, which are muscular bags of jelly, sieve water through fine sheets of mucus as they pump themselves along. Pteropods, or sea butterflies, are planktonic snails which either trap plant cells on their mucus-covered wings or produce extensive sheets of mucus which entangle any plant cell that touches them. Most extraordinary is a little tadpole-like creature, *Oikipleura*, that secretes a mucus house around itself which also acts as a sieve; every so often it eats its house, with its catch, and then makes another house.

All the herbivores provide food for a host of predators, some of which are themselves planktonic, such as the long, transparent arrow worms with their relatively huge, clawed mouthparts, and the tiny jellyfish with fragile-looking tentacles armed with stinging cells. Their relatives, the siphonophores, are colonies made up of interdependent organisms, each of which is specialized for a particular function, such as feeding, reproduction, swimming or defence. The most complicated ones have gas floats which contain high concentrations of carbon monoxide. The best known of these is the Portuguese man-o'-war, which floats on the surface and has such powerful stinging tentacles that it can paralyse a swimmer. Another is the sailor-by-the-wind *Velella*, which is washed up on the shores of western Europe after strong westerly gales.

Goose barnacles also get washed up attached to pieces of wood and old fishing floats. For many centuries these extraordinary animals were thought to develop into barnacle geese. Indeed, even when this myth had been dispelled, the animals were erroneously considered to be molluscs until John Vaughan Thompson, an army surgeon, established from examinations of their planktonic larval stages that they were crustaceans. Goose barnacles take only 14 days to grow from 2mm-long larvae to adults capable of producing more larvae. They use their legs as a trap which is sprung when any prey – which ranges from microscopic plants to quite large animals—touches one of the long, fine bristles, or setae.

One amazing feature of plankton is that a net towed at the sea surface by day catches little compared with one towed at night. Many planktonic animals swim down hundreds of metres each day at dawn and back up to the surface at night. Scientists still cannot agree as to exactly why this occurs.

PROJECTS

① In the spring, make your own plankton net using an old nylon stocking, or one leg of a pair of tights (preferably with no large holes!), minus the foot. Sew the top on to a rigid circle of wire attached to a long pole which will not bend. Pull the other, open end over the top of a

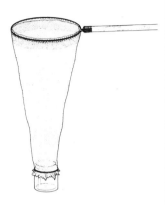

clear glass jam jar, securing it tightly with a strong rubber band.

Take a plankton sample by lowering the net from the end of a pier. The net will remain collapsed unless a flow of water passes through it. This can be achieved either by walking along the end of the pier so the net is towed through the water, or, if the tide is flowing, by holding the open end of the net against the direction of flow.

Gently remove the net from the jar without spilling the contents. Cut out a piece of black paper big enough to cover one side of the jar from top to bottom. Either fix the paper to the

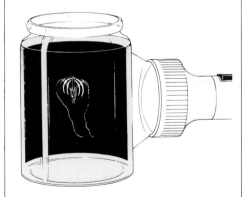

outside of the jar with adhesive tape, or gently slide it inside the glass. Shine a torch through the side of the jar so that the transparent plankton show up against the black paper. Use a hand lens to see the shapes of the smaller plankton.

FISH

Some marine fish have highly specialized methods of moving, feeding and reproducing. This results in some bizarre variations on the basic spindle fish-shape.

The most primitive fish which live in the sea are the jawless lampreys and hagfish which feed using their sucking mouths, lampreys living as parasites on live fish, and hagfish scavenging on dead and dying fish. Most marine fish, however, fall into one of two divisions based on their type of skeleton. The Chondrichthyes (Elasmobranchs), which include the sharks, skates and rays, have a skeleton made of cartilage; the Osteichthyes (Teleosts), which include the vast majority of fishes in both fresh water and the sea, have a bony skeleton. Other features of the Chondrichthyes include a tough, leathery skin, often covered with tooth-like denticles; separate gill openings (5–7 in sharks and 5 in rays) without a gill cover; rigid, paired fins which cannot be folded back alongside the body; and an asymmetrical tail fin with a larger upper lobe. Bony fish have a scaly skin; a gill cover, or operculum, protecting the gills on each side; and highly mobile paired fins known as the pectorals (front) and pelvics (hind). The eggs of most bony fish are fertilized externally in the sea, whereas the eggs of skates and rays are fertilized inside the female fish by the male gripping the female with his modified pelvic fins, or claspers.

The particular life style of a fish species is closely linked to its body shape. Bottom dwellers may be flattened so that their bodies blend in with the bottom contours. Flatfish such as turbot, brill, plaice, flounder, sole and dab, which are flattened from side to side, lie on one side with both their eyes uppermost (pp. 128–9), whereas rays, skates and the angler fish, which are flattened from top to bottom, lie on their bellies. The elongated bodies of conger eels and moray eels enable them to crawl over the sea bed and also to wriggle into rock crevices.

Fish which live in mid-water are built more for speed and have beautifully streamlined bodies. The stiff body of the mackerel helps the fish to store energy as it beats its tail. As the body flexes down one side, the muscles are stretched like elastic, so on the next stroke they spring back as well as contract. The mackerel also has a red muscle block running down the length of the body. Rich in haemoglobin, the

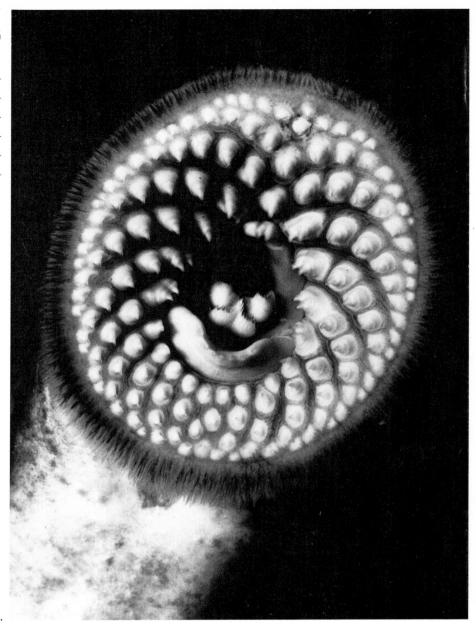

▲The sucking mouth of the parasitic sea lamprey *Petromyzon marinus* is highly adapted for clinging and feeding on the flesh of live fish.

▶A plaice *Pleuronectes platessa* head-on, showing the asymmetrical mouth, used for feeding on bottom fauna, and two eyes on the upper (right) side of the body.

▶A conger eel *Conger conger* snakes its way through the shallows of an eel grass bed during low tide.

◄The sea horse *Hippocampus* sp uses its prehensile tail for clinging on to weeds.

▲A red mullet *Mullus surmuletus* uses its pair of long chin barbels for feeding on the bottom.

▼Behind the eyes of this thornback ray or roker *Raja clavata* can be seen a pair of spiracles which are

used, when the ray is resting on the bottom, for drawing in a fresh supply of sea water over the gills.

danger and also by presenting the predator with a confusing, swirling mass of targets which are difficult to keep in focus. A shoal of herring will also sweep plankton from the water more efficiently than a scattered mass of fish feeding at random. Shoaling is often associated with breeding and this is when fishermen find herring and cod most easy to locate and catch. Off the Norwegian coast herring spawn from February to April, the eggs sinking to the bottom where they stick to rocks and shells. In a single year, as many as one to three million tonnes of eggs are laid in Norwegian waters. Cod lay buoyant eggs that float to the surface, where they hatch into larvae which live as part of the plankton for several weeks. Shore fish are more territorial; for instance, a blenny will drive other blennies out of its 'own' pool, partly to preserve its food supply but also to protect its eggs. Ways in which other shore fish guard their eggs or offspring are described on pp. 144–5. Some Chondrichthyes, such as the smooth hound and the tope, produce ten to twelve live young. A thresher shark, which reaches sexual maturity at a length of 4 metres, produces two to four young which are already 1.2–1.5 metres long when born. The largest fish in European waters is the basking shark, which, although it grows up to 12 metres long and weighs 3,500–4,000 kg, is not a fierce, fast-swimming carnivore, but a sluggish, plankton filter-feeder. The female gives birth to one or two young after a gestation period of two years. This contrasts greatly with the huge, comical-looking sunfish, a bony fish which can produce 300 million eggs. Not all Chondrichthyes, however, are viviparous, and the mermaids' purses of skate or lesser-spotted dogfish are by no means an uncommon sight on the shore (pp. 144–5).

Besides their eyes, fish have a variety of sense organs. Most important is the lateral line, a row of receptors usually in a canal which runs down the midline of a fish's flank. It senses low-frequency vibrations made by an approaching predator or a potential meal. Fish also have elaborate tasting and smelling senses. Cod and ling use the barbels projecting down from the lower jaw to locate food, while gurnards use the separate fin rays of their pectoral fins as finger-like probes for walking across the sea bed and finding their food, which consists of small fish and crabs.

Compared with the many commercially important bony fish, sharks and rays are less important as a source of food for man, although some can be seen in European fish-markets and dried shark fins are made into soups in the Far East. Shark skin is also used as a source of ornamental leather, which is known as shagreen.

red blood pigment that carries oxygen, the muscle has a special counter-flow blood supply so that it can conserve the heat which it generates by working. Like car engines, muscles work more powerfully when they are warm. Both mackerel and sharks have to swim continuously to maintain their position in the water because they do not have gas-filled swim-bladders and so are denser than water.

The swim-bladder plays a very important role in the way a John Dory feeds, allowing it to hang motionless in the water. The John Dory stalks its prey very slowly, presenting its highly laterally compressed body head-on to its prey so that it appears as a thin vertical profile. The prey is sighted by the huge eyes and sucked up by the protrusible jaws extended forwards as a tube.

Pipe-fishes have an eel-like body with the jaws modified into a pipette for sucking up small plankton. Sea-horses have similar jaws, but the body is armour-plated and the tail is prehensile

for clinging on to seaweeds and eel-grass. Several fishes living in the surf zone, such as gobies and lumpsuckers, have developed sucker-like fins for clinging safely to stones without the danger of getting dashed against the rocks.

Fins are also modified for various uses; for example, the irregular outline of its laterally placed fins helps to camouflage the angler fish as it lies on the bottom. At the front of the head is a large, mobile spine with a fleshy tip which is used, rather as a fisherman uses his baited line, to entice other fish to within striking distance of its huge, gaping, crescent-shaped jaws—jaws that are quite different from the small ones of the grey mullet, which are modified for filtering very tiny animals out of fine mud.

Fish behaviour varies with life style. Angler fish are solitary, whereas mullet feed in shoals. Shoals give protection against predators both by increasing the number of eyes on the look-out for

165

FISHING AND FARMING

Shell middens all over the world prove that man has collected food from the sea for thousands of years. He now uses highly sophisticated techniques for harvesting shellfish and fish, and also farms some species.

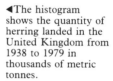

▶Otter trawl contents strewn on to the deck of a trawler include red gurnard *Trigla cuculus* and pout *Gadus luscus*.

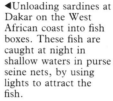

◀The muscular tails of Norway lobsters *Nephrops norvegicus*, packed in ice coming ashore from a trawler at Höfn in Iceland, will end up as scampi.

◀Unloading sardines at Dakar on the West African coast into fish boxes. These fish are caught at night in shallow waters in purse seine nets, by using lights to attract the fish.

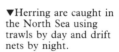

◀The histogram shows the quantity of herring landed in the United Kingdom from 1938 to 1979 in thousands of metric tonnes.

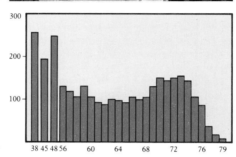

▼Herring are caught in the North Sea using trawls by day and drift nets by night.

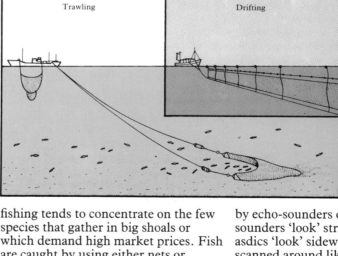
Trawling — Drifting

Primitive man was originally a hunter-gatherer; the men went out to hunt for fish or game, while the women grubbed for roots or collected fruit. Today, our sole major method of hunting for food is by fishing.

Early fishing methods

Originally, man collected individual crabs, mussels, limpets and fish during low tide. At a later stage, he built stone walls and traps for retaining plentiful seafood when the tide receded. During the Stone Age, fish hooks were being made from stone, bone and horn as well as wood. By the time of the Bronze Age, boat building was developing, which enabled the fishing to be undertaken away from the immediate shore. As many as 3,000 years ago, the Phoenicians had learnt to preserve fish in salt, which enabled them to develop a prosperous salt fish trade.

Fisheries became such a vital part of the commerce of many coastal European towns that they incorporated fish into their coats of arms. Fishing ports on the east coast of England began catching herring as far back as AD 1000. Herring was also caught in Scandinavian countries, where the practice of hanging stockfish (usually cod) to dry on outside racks has been carried on since the early Middle Ages. Cod liver oil was originally used to preserve leather; but it was not until 1775 that it was known to prevent rickets. In 1685 the cod fishery failed completely in the Faroes, as a result of abnormally low sea-water temperatures, which also resulted in Iceland being completely surrounded by ice throughout the year.

Modern fishing methods

Although the principles of modern-day fishing are based on primitive techniques, the aids have become very sophisticated. Marketing and economics mean that commercial fishing tends to concentrate on the few species that gather in big shoals or which demand high market prices. Fish are caught by using either nets or hooks, with variations on these themes. Nets are used in mid-water to collect shoaling species such as herring and mackerel. The old technique of drift-netting has been replaced either by big trawls in which massive 'doors' are used to keep the net sides wide apart, with floats to keep the head of the net up and weights to keep the front down; or by purse seine nets, which are shot in a circle around a shoal, and the bottom of the net enclosed by pursing it with a draw-rope. The shoals are first located by echo-sounders or asdics: echo-sounders 'look' straight down, whereas asdics 'look' sideways and can be scanned around like radar. In some shallow water areas, large permanent net traps are used for catching herring, mackerel and eels.

Long-lining is a technique used for catching big oceanic fishes such as tuna, swordfish, marlin and various sharks. Many miles of line are buoyed up near the surface, and vertical snoods, 2–3 metres long, are clipped on at 20–30-metre intervals with a hook on their ends.

Trolling, which involves using lures or spinners on the end of a big rod towed at a high speed from a boat, is

◄Live oysters *Osrea edulis* being cleaned overnight in trays in an oysterage, prior to marketing.

▼Wooden racks festooned with headless cod, hung up to dry by the sun and the wind, are a common sight in Iceland.

like jaws) that live in mid-water depths of 1,000 metres or more.

Crustaceans and molluscs

Crabs and lobsters are caught on a commercial scale in baited pots, the annual European catch of lobsters being 2,000 tonnes while the annual British catch of edible crabs alone is 5,000–6,000 tonnes. Shrimps are caught in trawls all round the European coast, and the larger Dublin Bay prawns (which resemble miniature lobsters in shape and are sold as scampi) are caught in both trawls and seine nets in European waters between Spain and Iceland. Both scallops and the smaller queen scallops are caught commercially, as well as oysters, and squid, which is a south European delicacy.

Over-exploitation

Attempts have been made to regulate the quantities of fish that are caught each year so that the greatest catch is taken without reducing the total stock. During the last World War, no fishing was carried out in the North Sea for several years. In the immediate post-war years, catches were exceptional. The fish bonanza attracted considerable investment in the fishing industry, so that not only did the size of the fishing fleets increase but the method of locating the fish became much more sophisticated.

The herring story exemplifies the subsequent depressed sequence of events. The herring has often played an important role in the development of various maritime nations in Europe. This fish spawns on banks where the eggs cover the sea bed as a complete carpet. The eggs hatch into planktonic larvae which drift inshore and develop in the shallows. Herring exists as distinct races, each of which spawns at different times of year. Once this was known, the herring fleets simply moved from one spawning ground to another. The development of factory fishing techniques in which shoals of young fish were pumped up, sieved off and turned into fish meal, resulted in such a disastrous reduction of the spawning stocks that in the late 1970s herring fishing was banned throughout EEC waters. In 1981, Britain endeavoured to maintain a ban on fishing the only remaining large herring stock which spawned off west Scotland, so as to allow it to build up to greater strengths. Other EEC countries insisted that fishing should restart.

Nearly all fish stocks in European waters suffer from chronic over-fishing, resulting in political strife such as the 'Cod War' between Britain and Iceland. As a means of conserving what was left of Iceland's major national resource—its cod stock—she extended her territorial limits unilaterally.

The reclamation of tidal flats, notably in Holland, and the pollution of estuaries, has reduced important nursery grounds for many commercial fish, for example the plaice. In an attempt to locate new fish stocks for exploitation, commercial trawls have been made to depths of 1,000 metres; vast shoals of blue whiting have been located in the Western Approaches; and, most recently, the economic viability of fishing krill—the shrimp-like crustacean food of baleen whales in the Southern Ocean (pp. 168–9)—has been investigated.

Farming the sea

Clearly, the best theoretical solution to the over-fishing problem is to farm the sea. Success, however, is dependent on gaining a measure of control over the environment. On land, forests can be cleared, the soil ploughed, and alien weeds or animal pests kept at bay; whereas the vast marine environment is much less predictable and accessible, and therefore less manageable.

In the Far East, coastal ponds are used to grow vegetarian fish, such as mullet, and omnivorous prawns. They are stocked with naturally occurring larvae which do not need supplementary feeding in the ponds. Attempts to grow live plaice from eggs in the warm water effluents of power stations have proved successful in obtaining small fish with a high chance of survival in the sea. Unfortunately, if these fish are returned to the sea, there is no way of stopping them swimming off into the territorial waters of other countries which have not contributed to their juvenile upkeep. Fish farming has therefore been concentrated on species which are not highly mobile and which take little effort to feed, or on those species which command very high market prices. In the rias of north Spain, numerous moored pontoons are used to support hanging ropes on which vast numbers of mussels are grown. Similarly, in the great bay at Arcachon near Bordeaux, the outgoing tide uncovers a great patchwork of wickerwork compounds in which oysters are grown. Both these farming techniques depend on natural spatfall of the tiny planktonic larvae.

A more complete farming technique is used to rear trout in the sheltered sea-water lochs of Scotland and the fjords of Norway. After the eggs are hatched, the young fish are reared in wire compounds, where they are acclimatized to sea water. They are fed either with pellets or with deep-frozen plankton, trawled up during the spring bloom. Marine farms, however, need very much more investment before man can harvest the seas at levels approaching the yields from even the poorest of mountain pastures.

mainly a sports fishing technique, but it is carried out on a commercial basis for high-value fish such as the blue-fin tuna. Small species of tuna, such as the skip-jack, are fished with rods fitted with unbaited and unbarbed hooks. Shoals are located, and by spraying the surface with a jet of water, a feeding frenzy is induced. The fish snap at anything. If they hit a hook, the fisherman strikes, flipping the fish inboard where it drops off the unbarbed hook, which is then flicked straight back into the water. Droplines are used for fishing bottom-living fishes and by Madeira fishermen for catching scabbard fish (long, slim fish with wolf-

MARINE MAMMALS

Seals, whales and sea cows are highly adapted for their life in the sea. Like all mammals, they are warm-blooded air-breathers, giving birth to live young, which are suckled.

The strangest seal in the north Atlantic is the walrus. In medieval times, the tusks of the male walrus were highly prized as ivory, and the herds were greatly reduced by hunting. Later, seals were hunted for their skins, particularly the cubs of harp seals, which are heavily culled in Canada, Greenland and Spitsbergen. Seal culls are now being advocated to control seal populations as a protective measure for inshore fisheries, especially salmon fisheries. Random killing of seal pups by fishermen has reduced the populations of the Mediterranean monk seal—the only seal to live in warm, tropical and sub-tropical waters—to dangerously low levels.

Since seals haul themselves out on to land or ice to breed, estimates of their populations can be made using aerial photography. Along the European coasts, white pups of the grey, or Atlantic, seals and common seals show up well against the darker ground, whereas pups of seals which haul themselves out to breed on ice are perfectly camouflaged. The only way these pups will show up on film is through the use of ultra-violet photography, when they appear black against the white ice.

Seals, like whales, conserve their body heat with a layer of fat or blubber, but, unlike whales, they also have a furry coat or pelt. The valuable pelts of the fur seals have resulted in the dramatic exploitation of these mammals by man. The hind pair of limbs of true seals is flipper-like and useless when they come out on land or on ice, whereas sea lions have hind limbs which are still useful on land. Most seals are fish eaters, but walruses use their long, sensitive whiskers to locate the bottom-living shellfish on which they feed.

Seals have large, well developed eyes and hunt mainly by sight. In contrast, whales have small eyes and hunt more by echo-location, sending out a series of sound waves which rebound from objects as echoes, enabling a whale to find food and communicate within the school. Associated with this method of hunting and communication, many

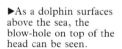
▲A dead fin whale being pulled ashore up to the flensing platform of a whaling station in Iceland.

►As a dolphin surfaces above the sea, the blow-hole on top of the head can be seen.

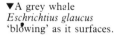
▼A grey whale *Eschrichtius glaucus* 'blowing' as it surfaces.

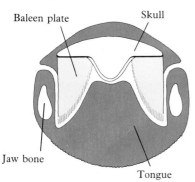

Baleen plate   Skull

Jaw bone

Tongue

▲Section through the head of a blue whale showing how the fringed baleen plates hang down vertically. Mouthfuls of seawater are taken in and squirted out of the sides of the mouth. The fringes of the baleen plates sieve off plankton, which is then swallowed.

▲A harp seal pup *Pagophilus groenlandicus* is born on ice.

▼Various stages in a marine food chain starting with phytoplankton and ending in a sperm whale. (Not to scale.)

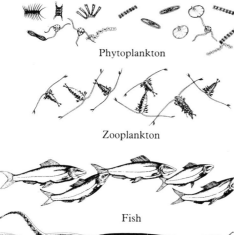

Phytoplankton

Zooplankton

Fish

Giant squid

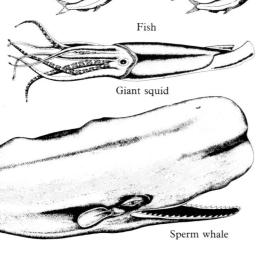

Sperm whale

whales have evolved a high intelligence. Dolphins, seals and even killer whales have been trained by man to perform feats in oceanaria.

Whales, in particular, are highly adapted to their aquatic life, with the pregnant cows giving birth to their calves in the sea. The forelimbs of a whale are modified into flippers, which are used as hydroplanes, the huge tail flukes giving the main propulsive stroke. The tail, which is flattened from above and below, is moved vertically up and down, unlike a fish's tail which is flattened from side to side and therefore moved sideways.

Whales have nostrils which open on the tops of their heads, so that they inhale and exhale, or 'blow', in the second or two when their head rises above the water. There are two main groups of whales: the baleen whales and the toothed whales. The first group has fringes of hair arranged along the horny baleen plates which hang vertically down at the sides of the mouth. When the whale takes a mouthful of sea water (5 tonnes at a time in the case of the blue whale) it forces it through the baleen with its tongue. The biggest blue whales weigh over 100 tonnes and the tongue alone is as heavy as an elephant. Plankton suspended in the water are filtered off by the baleen and swallowed. Blue whales eat mostly shrimp-like krill; sei whales have finer hairs and eat smaller plankton; fin and minke whales take fish as well as krill. Whales feed on any animals that concentrate in big enough shoals. The most highly modified whales are the right whales, whose baleen plates may be over 4 metres long. Right whales are so-called because they float when killed, so they were the best, or right ones, for the early whalers to kill.

The Biscayan right whale was hunted by the Basques from open boats during the eighteenth and nineteenth centuries before it was exterminated by them. Many other baleen whales are now threatened by over-exploitation, but pressure from conservationists may save humpbacks, fin, sei and, it is hoped, the largest of all, the blue whale.

Toothed whales are fast-swimming carnivores feeding on fish and squid. The largest of these is the sperm whale, object of the Yankee whaling fleets that inspired Melville's classic novel *Moby Dick*. The big sperm bulls grow up to 15 metres in length, and are very aggressive in the breeding season when they assemble the smaller (10 metres long) cows into harems. Sperm whales feed mostly on squid, and their stomach contents are the main sources of material for scientists studying giant squids. These whales hold the deep-diving record set by a bull which was found entangled in a submarine cable at a depth of 1,200 metres, and

circumstantial evidence suggests that they can dive to at least 2,000 metres. In the Faroes early man learnt how to drive schools of pilot whales ashore in shallow bays. Rarely seen in European waters are narwhals, with their long, tusk-like tooth that probably gave rise to the unicorn myth. The fiercest hunter among whales is the killer whale, packs of which hunt down seals, birds and even other whales, particularly dolphins. They are not an uncommon sight in the north Atlantic Most familiar are the dolphins, whose intelligence makes them firm favourites at oceanaria.

Whales tend to make spectacular migrations. They move polewards in the spring to feed on the vast populations of plankton and fish that build up in response to the spring bloom of phytoplankton. Later in the year, when the rich feeding comes to an end, they move back into warmer subtropical and tropical waters, often not feeding but merely surviving on their food reserves. This tends to be when they breed, so that while the newly born calves are growing they do not lose too much heat. Sperm whales have the richest milk known; a third of it is fat, and the calves put on over 3 kg of weight per day.

The latest threat to the existence of baleen whales is the rapid growth in the exploitation of krill in the Antarctic. Seals are believed to consume annually 64 million tonnes, whales 43 million, and the penguin chicks of South Georgia alone, another million. If man starts to take huge catches of krill without care, a massive ecological disaster could ensue.

Whales, in general, are long-lived, with a life span of up to a century. In order that changes in whale populations can be detected, much more information on the biology both of individuals and of populations is badly needed.

Whales are aged by means of growth rings laid down in their waxy ear plugs, in the baleen plates, and in the dentine layers of the teeth of toothed whales. Unlike seals, which haul themselves out to breed, most whale populations cannot be counted from aerial photographs. Metal tags used to mark individual whales can be recovered only when the whale is processed by the whaling industry. Hence, conservationists have a dilemma that unless whales are exploited they cannot get the information required for proper scientific management of whale populations.

Few animals have received more attention from ardent conservationists in recent years than marine mammals—especially whales—yet many of these activists have never seen, in the wild, the animals that concern them so much.

MARINE POLLUTION

Marine life is threatened by not only major oil spillages, but also the disposal of domestic, agricultural and industrial waste, including the discharge of pesticides, warm water and heavy metals.

The sea has long been regarded as a bottomless dustbin into which man can throw all his rubbish in the belief that it will disappear. As the population throughout Europe has increased, and the communities have grown more affluent, it has become clear that the sea cannot absorb all our rubbish.

Sewage

The most obvious sign of pollution is that of untreated sewage, an unpleasant discovery made by the bather who comes face to face with a sewer outflow at low tide. For the swimmer, there is the risk of contracting infectious intestinal diseases such as typhoid, paratyphoid, polio or diarrhoea from swallowing contaminated sea water. The EEC directive for acceptable bathing water (which will have to be met in 1985) sets specific limits for the number of intestinal bacteria contained in 100 ml of water. In the late 1970s, only twenty-five British beaches which were visited by large numbers of bathers met this standard.

For the shore inhabitants, the effects of sewage are different. Flocks of gulls feeding around outfalls show that some animals derive real benefit, and anglers often report their best catches in these sites. But this input of rich organic matter can increase the turbidity of the water and so reduce the maximum depth at which seaweeds can grow. It also increases the biochemical oxygen demand (BOD) of the water, as bacteria require oxygen to break down the organic substances in the sewage. In some habitats, notably muddy shores and estuaries, it greatly increases the possibility that the mud will become totally devoid of oxygen. Under these anoxic conditions it becomes black and very smelly, stinking of hydrogen sulphide. As in freshwater muds, this black layer is almost totally lifeless, for there are few species that can exploit such anaerobic conditions.

Oil spillages

It is interesting that after the *Amoco Cadiz* incident in 1978, where the oil

▲ Oil tankers re-fuelling at Sullom Voe Terminal at Shetland, the site of the major oil spill from *Esso Bernicia* in December 1978.

▶ Oil rigs, such as this one in the North Sea Ekofisk oil field, burning off propane gas, are another potential source of oil pollution of the sea.

◀ An oiled immature razorbill *Alca torda*, which has been beached on the shore, is attempting to walk down the beach at Great Yarmouth, on the east coast of Britain, in February 1981.

▼ During low tide, the effluent from a sewage outlet on the south coast of Britain is clearly visible. Here, seabirds are attracted in to feed, while bathers are deterred from swimming.

◀ Instead of dumping his rubbish inland, man has thrown it from the cliff-tops of a small island. The action of the sea has accumulated a cooker, a lawn-mower, bicycles and other eyesores on the beach overlooked by visiting tourists.

spill killed all the animals inhabiting sandy and muddy shores, the first invaders were the same worms as those which exploit shores heavily polluted with sewage. Oil pollution, whether on the chronic level caused by minor spills in the vicinity of oil terminals, or on the grand disaster scale as produced by the *Torrey Canyon* (1967) and *Amoco Cadiz* spillages, and the Ekofisk blow-out, creates most publicity when seabirds are badly oiled. The oil clogs their feathers and destroys their insulating effect. In attempting to preen the feathers, the birds swallow quantities of the toxic oil, and even after careful cleaning the majority die. Diving birds mistake slicks for fish shoals, with disastrous results. Various species of the auk family (puffins, razorbills and guillemots), which gather in vast rafts prior to breeding and afterwards to moult, are particularly vulnerable to oil contamination.

But these heart-rending effects on seabirds are only part of the story. On any shore suffering from heavy oil pollution, complete communities of animals and plants may be smothered and die. Indeed, considerable damage may occur to the commercially exploited shellfish, such as oysters, which become tainted with oil, though these can usually be cleaned up by keeping them in non-polluted water for a few weeks.

Although some of the components of oil are gradually biodegraded, the major natural detoxifying mechanism is by dispersion. However, in most cases, and particularly on sea shores, both natural dispersal and biodegradation are intolerably slow and some other method must be used to reduce the level of contamination. Various types of detergent are used for this purpose but they create an added hazard. If the wrong detergent is poured directly on to beaches, it will kill intertidal and sublittoral life as effectively as the oil. Since the *Torrey Canyon* disaster, much research has been carried out to produce less toxic detergents; but in the mid-1960s it was common to see huge beds of gaping mussel shells and conspicuous strandlines of rotting limpets and other molluscs on shores when detergents had been used. The detergent also seemed to cause crabs to lose many of their legs and to kill razor shells below the low tide mark.

After marine life has been destroyed, recovery by means of recruitment from neighbouring shores may be quite fast—within three years—for species with planktonic larvae. Reinvasion takes much longer, however, for the species which lack a planktonic dispersal phase in their life history.

Toxic waste

More insidious forms of pollution are those of heavy metals and organic compounds such as DDT and polychlorinated biphenyls (PCBs). These pass into the marine environment by way of river run-off, or through the atmosphere in the form of dust, or in rain water. Insecticides like DDT, and lead compounds added to petrol as an anti-knock agent, enter the sea from the air. Emission of smoke from stacks of smelting works and other industrial complexes are monitored, but they still release significant quantities of toxic compounds into the atmosphere.

Discharges of chemical wastes into rivers and estuaries have been known to cause numerous deaths in man, notably in Japan where both mercury and cadmium poisoning have killed people eating fish and shellfish from heavily polluted water. Many heavy metals are rapidly absorbed into sediment particles suspended in the water, which settle out at slack tide. Thus, mercury, copper and many of the radioactive isotopes discharged from the nuclear waste processing plants of Windscale, in Cumbria, England, and Le Harge, in Brittany, France, are removed from the immediate environment. However, some heavy metals, such as cadmium, are not removed by sediments and tend to stay dissolved in the sea water, where they are more available for animals or seaweeds to absorb. The Windscale plant provides the most detailed example of a marine monitoring programme anywhere in the world.

The really dangerous pollutants are those that are accumulated by animals, especially if they concentrate them from their food. DDT and PCBs are both accumulated in this way, with the result that the top predators in the ecological pyramid accumulate vast quantities in their body tissues—quantities that are either directly toxic or prevent them from reproducing successfully. Evidence suggests that, whereas mercury and lead may be concentrated up the food chain (but fortunately tend to get removed from the marine environment by other processes), the vast majority of radioactive elements discharged from nuclear plants, such as polonium, uranium and thorium, are not accumulated in the food chain.

Heating up the water

Man's use of energy adds two other pollutants into the ecosystem: heat and carbon dioxide. The cooling water used by coastal and estuarine power stations is emitted as warm water, as much as 12°C above the ambient water temperature. This warm water is lethal to organisms living adjacent to the outfall and it also reduces the oxygen-carrying capacity of the water. However, it is rapidly cooled down as it enters the mass of cooler water, and, compared with much smaller freshwater bodies, heated effluents entering the sea have a minimal effect.

Increased outputs of carbon dioxide are potentially more frightening. The long-term consequences of higher carbon dioxide levels in the atmosphere are controversial. However, most meteorologists are agreed that the initial effect would be a small rise in atmospheric temperature throughout the world as a result of additional insulation provided by the carbon dioxide against overnight heat loss. Levels of the gas, as measured over the relatively isolated island of Hawaii, have doubled in the last fifty to sixty years. If this increase continues, not only will the polar ice caps begin to melt, causing a substantial rise in sea level with subsequent flooding of low-level ground, but there would also be a noticeable shift in the climatic zones of the earth, resulting in major environmental changes on land, in fresh waters and in the sea.

MAN'S IMPACT ON THE SEA

As a result of man's activities, marine habitats have been altered both below low water mark and higher up the shore. Such changes may allow a few adaptable species to thrive, but they often tend to reduce the natural diversity of species.

Settlements grew up along the coast in places with sheltered waters, particularly in estuaries. Coastal fortifications were then built as protection against the violence of the sea and attack by seafaring raiders such as the Vikings. More seaworthy boats led to longer journeys and the dawn of trading between nations. Even in the Stone Age, tools were traded across the English Channel.

In recent years, marine pollutants have been responsible for major impacts on life in coastal waters, as exemplified on the previous spread (pp. 170–1). The sophistication of fishing techniques, including the use of sonar devices to locate fish shoals accurately, has led to the over-fishing of commercial species (pp. 166–7). Other major areas in which man's impact has had effects on marine life have resulted from building coastal structures, dredging the sea bed, introducing alien species, and pressure from seaside visitors.

Harbours and marinas

The building of harbours, and coastal protection in general, modifies natural habitats quite substantially. Coastal protection ranges from the building of sea walls to wooden or concrete groynes constructed to help stop the longshore drift of sands and shingle (pp. 118–19). These structures provide new microhabitats in areas which often lack environmental diversity. Seaweeds, barnacles and mussels find hard substrates on which they can settle and attach themselves, while the wood-boring gribble gains a more extensive area in which to bore.

In harbours, there is usually an increase in the pollution level, which will restrict the range of species, but those animals that are able to withstand the pollutants can abound. Floating pontoons, or locks which maintain a

▲A sonograph, or sound picture, made by a short-range side-scan sonar. Sound waves were bounced off the sandy sea floor to produce a shadow picture of trawl marks at a depth of about 40 metres. Taken in the North Sea, this picture was made by the Geology Group of the Institute of Oceanographic Sciences working from RRS *Discovery*. 12 pairs of trawl marks, each less than 15 cm deep, can be seen criss-crossing over the sea floor.

◄Colonies of this attractive sublittoral hydroid *Tubularia indivisa* grow attached to pontoons in Southampton Water.

▲Completed in 1978, Brighton Marina is the largest in Europe, and has moorings for over 2,000 yachts.

◄Exposed at low tide, the base of the piles supporting a jetty in Guernsey provide habitats for a variety of sessile marine organisms, including sea squirts *Ascidia mentula* and sponges *Grantia compressa*.

constant level inside a harbour, can create conditions where sublittoral animals which normally live deep down can live just below the sea surface, and so are more easily accessible to naturalists and marine biologists.

For many years, Swansea docks in south Wales were inhabited by a number of exotic warm-water species of worms and other invertebrates which had been introduced by ships steaming in from warmer waters. These aliens arrived either on the hulls of the ships or in their bilge water, and were able to survive because the dock was warmed by a power station outflow. Now the dock is no longer warmed, most of these exotic species have died out.

The building of marinas destroys the original coastal habitat and replaces it with extensive concrete walls. The 51-hectare marina at Brighton in the south of England has over 2,000 moorings, making it the largest yacht harbour in Europe and the second largest in the world. Building began in 1971 and was not completed until early 1978. The two curving breakwater arms enclose the outer harbour which is tidal. Leading from this is a 100-metre-long lock which passes into the 2.5-metre-deep inner harbour. This marina provides shelter not only for boats, but also for many commercial marine invertebrates, for its breakwaters are made out of hollow concrete drums, known as caissons, each of which weighs 600 tonnes. The insides of these caissons are

▼Japweed *Sargassum muticum* is an alien seaweed which grows rapidly, soon swamping native seaweeds such as this serrated wrack *Fucus serratus*. Here it is growing at Bembridge on the Isle of Wight, where it was first discovered in 1973.

used for farming oysters, rainbow trout and lobsters. Promenades along the top of the breakwaters allow fishermen easy access to the outer sea.

Dredging

The need for shipping to have clear channels means that dredging is an activity centred on ports. The dredge slurry is either taken in hopper barges which dump their unwanted cargoes in deep water, or piped ashore to help with land reclamation. These activities affect not only the animals that are dug up from the seabed and dumped in a foreign habitat, but also the communities that live down-current of the dredging operations. Filter-feeders need a modicum of suspended material in the water on which to feed, but a dense cloud of clay and fine sand clogs their filters and their gills.

Dredging is also carried out to supply gravel for building, particularly now that gravel deposits on land are being depleted. Gravel dredging in the sea is carried out in two ways. One method is to take samples until an area of the desired grade of gravel is found; this is then removed using a suction dredge. The alternative method, known as ribbon dredging, creates a long furrow across the sea bed as the dredger drifts with the current. The gravel deposits in the English Channel and the North Sea are old river terraces and glacial moraines laid down during the Ice Ages when these areas were dry land, and so they will not be replaced with more gravel. Consequently, the dredge scour marks remain as sea bed structures which create problems for fishermen trawling these grounds. Gravel excavation can also cause erosion elsewhere on the sea bed as sediments are carried along to fill in the dredging holes.

The laying of pipelines to bring oil and natural gas ashore creates only temporary environmental disturbance since the pipes are normally laid in trenches which are then filled in. However, on several parts of the North Sea bed, systems of sand waves occur. These are analogous to sand dunes on the shore, but are moved along by the prevailing underwater currents. Pipelines which have to be laid through sand wave systems are sometimes uncovered by the waves moving through them; consequently they become vulnerable to damage by storms and fishing gear. Exposed pipelines are a bonus, however, for animals which normally inhabit rocky bottoms, since they present a hard surface on which they can settle.

Alien species

Ever since man moved from one country to another by boat he has been responsible – both unintentionally and

deliberately – for the introduction of exotic species into coastal waters. Some introductions, such as the Australasian barnacle which was introduced to British waters via ships' hulls during the Second World War, compete for rock space on which to settle with the native species. A much more recent competitor to the British shore scene is the brown seaweed known as japweed, which was first recorded in February 1973 on the Isle of Wight. This seaweed not only grows very rapidly, thereby swamping the native brown wracks, but it also has a rapid means of dispersal into new waters since small fragments breaking off the parent plant will float away and develop sex organs. This ability to fragment so easily means that it not infrequently fouls up outboard motors. Japweed also displaces eel grass which can form quite extensive marine meadows on some estuarine shores, where it is the prime food source for overwintering brent geese (pp. 114–15).

When American oysters were introduced to Britain, two oyster pests came too; these were the oyster drill and the slipper limpet (pp. 124–5). The oyster drill is a direct predator which attacks oysters by boring through their shells; whereas the slipper limpet is a more efficient filter-feeder, thus competing directly with the oysters both for food and for bottom space.

Seaside visitors

Access to the coast has now been made easier throughout Europe by improved roads, and permanent caravan sites mean that the coastline is under increasing pressure for recreation. Disturbance to nesting birds – particularly tern colonies – can create a minor ecological disaster by allowing predators access to the eggs and chicks (pp. 130–1). Repeated trampling can destroy the ground cover of plants, so that the bared cliff-tops erode away; also, blow-outs occur in sand dunes on which stabilizing plants, such as marram grass, are prevented from establishing themselves (pp. 126–7).

There are not a great many sandy beaches where it is possible to take a walk during a low tide in mid-summer and return without seeing any footprints other than your own. It is well known that only the most ardent walkers venture far from their car, so that if coastal car parks are restricted in size, they will automatically limit the number of visitors per day, which will markedly decrease the annual pressure on sensitive sites due to trampling.

No marine habitats are safe from the threat of pressure by man. Even rocky shores, which are usually impractical for development, can suffer from pollution and over-collection by repeated visits from educational groups. 173

MAN'S IMPACT ON INLAND WATERS

Man has consistently used inland waters for his own ends, damming, diverting, straightening and draining them to supply his needs. How does wildlife adapt to this management?

Inland water represents one part of the water cycle, where water is either on its way to the sea or being recycled through evaporation from the water surface, from plants or from the earth. Left to itself, the water on its journey creates a variety of habitats, each with its own community of interdependent species. But as the human population has increased, man has needed more farming land and more water. Both of these needs require intensive management of wetlands and freshwater resources, the object being to maintain a system that only stores water or supports the species that are useful to man.

Interfering with the complex natural system often has unforeseen results, however. The hill farmer who ploughs more hectares for arable land increases potential soil erosion by storm water; increases the rate of evaporation from the soil surface; and lowers the water table, which may also affect the spring-line on which he is dependent for water. The lowland farmer who needs more arable land must drain, divert and channel the water so that the land will dry out before he can crop it. But as it dries it shrinks, and as the land level is lowered the drainage levels change and the possibility of the incursion of sea water increases in coastal areas. In other areas irrigation may bring into cultivation land that, although rich in native species, produced nothing of food value for man. But maintenance of an irrigation system is costly, for the water is constantly removing banks and silting up the channels. Maintenance also reduces the habitats available for aquatic species. In large areas of the world rice is a staple food, extensively cultivated in Asia by an elaborate system of irrigating whole fields with standing water. It is also grown in Europe, in the Po valley in Italy and in the Algarve in Portugal.

Man strives to stabilize the water system for his own ends; but achieving

◀This river has been straightened here to allow faster flow and reduce risk of flooding. It is also regularly cleared of the abundant fringe and floating water plants.

▶*Crassula helmsii*, a recent introduction from Australia, is establishing itself beside a pond in the south of England.

▶The Asiatic grass carp *Ctenopharyngodon idella* may be suitable for introduction to control weed growth in dykes and ditches.

◀Rice paddies in the Algarve, in Portugal, where water levels are artificially maintained with an elaborate system of drains, channels and levels.

▼A village pond in winter is covered with ice, except where the ducks are preventing it from freezing. The filling-in of village ponds has restricted the freshwater habitats for many common species, such as the frog and the duck.

a balance between drainage, erosion and flooding requires a thorough understanding of each water system. Straightening and deepening the channels of lowland rivers is an essential part of flood prevention, yet straightening in one part may create flooding in another. Each system of dykes and drains must send the water in one direction, and the channel must be kept clear, thereby reducing the fringe vegetation for nesting birds. Drainage may often require an elaborate system of water pumps and raised banks. Although some drainage of wetlands occurred in medieval times, particularly under the monasteries, today, owing to mechanization, it is carried out over extensive areas. In many countries government money is available as grants to aid drainage improvements, which have a far-reaching effect on wildlife (p. 176).

Damming water courses is also a job for the water authorities in the provision of water or power for the ever increasing populations and their needs (pp. 60–1). This has been done on a much smaller scale for centuries by the creation of village ponds, and of weirs for the mill or other water-powered machinery. As people have no longer needed to water their stock or use a water wheel, these have fallen into disuse, the pond filling in with silt and rubbish and the stream by-passing the weir. In 1974 a Save the Village Pond Campaign was launched in Britain, during which over 1,000 ponds were restored – many by the British Trust for Conservation Volunteers, so that today they are again part of the village landscape, just as pools, streams and lakes have always played a central role in landscape gardening.

New places for old species

Although we are losing many inland water sites, we are gaining others through man's alterations of the drainage patterns. Vast reservoirs of fresh water are created, and are used by numerous wild species of plants and animals, particularly birds. Extraction of gravel or clay from old river beds has created inland lakes and ponds that are soon colonized by many wildfowl, and if managed properly should attract even more (p. 178). Mineral or other extractions from the earth may lead to subsidence and new water bodies, or old peat extractions may flood (pp. 176–7). A number of these places are now valuable reserves, where the environment is managed to attract as many aquatic and wetland species as possible.

Drainage may remove the wetland habitat, but the dykes and drains that carry permanent water often become a reservoir of wetland plant species, such as reedmace, common reed, bur reeds and frogbit, all of which may be abundant in the new habitat. Others have found a niche in particular conditions, such as those found in abandoned canals (pp. 98–9).

There are species, however, that have not found a niche in the new habitats and have become increasingly rare, such as the greater spearwort and the marsh helleborine, and birds such as the bittern, the black-tailed godwit, and many others. A vanishing habitat may be one cause of these losses, but in a few species, such as the cranberry, a changing climate has contributed to its disappearance in many southern localities (pp. 176–7).

Introductions

Man has had another striking effect on wetland wildlife by introducing alien species. If the climate and conditions are suitable for growth and breeding, these introductions may undergo a population explosion when removed from their native ecosystem with its associated competitors and predators. The Canadian pondweed (pp. 20–1), the coypu, the mink, the American freshwater shrimp *Crangonyx* and, more recently, the Australian aquatic plant *Crassula helmsii*, have become established in Britain and mainland Europe.

The Asiatic grass carp is a fast-growing, herbivorous fish which is being considered in Britain for the biological control of water weeds in enclosed waters in the East Anglian fens and the Somerset Levels, where mechanical diggers and herbicides are at present used to keep the drainage channels clear. As with any deliberate introduction, the impact of the species on the native population cannot be predicted with certainty. The advantage of introducing the grass carp into British waters is that it will reproduce only if there is a marked rise in the water temperature as well as in the water level; these conditions would rarely, if ever, occur in Britain, and the carp can therefore be kept under control. Experimental work carried out with native bream and grass carp has shown that the bream grow best in pools where grass carp are also present. It is possible that the bream are feeding on the faeces of the grass carp.

Man as predator and polluter

The fly fisherman who uses all his cunning to catch a salmon along a stretch of river is behaving as does any top carnivore at the end of a long chain of events and food resources affecting different consumer levels. At this level he necessarily competes with the other top carnivores, such as the otter. Since prehistoric times man has attempted to eliminate other carnivores, both by hunting them and by the removal of their habitats. In this way the beaver has gone from most of lowland Europe, and the otter is declining fast, except along uninhabited northern coasts. This picture is complicated by the introduction of another carnivore, the mink, which competes with the otter for food. But there are other side effects to man's predatory fishing habits.

At the end of the 1970s, a field project undertaken by the Young Ornithologists' Club in Britain revealed an unexpected link between coarse fishing and the drastic decline of the mute swan. Members were asked to collect lengths of discarded fishing line, hooks and lead shot (used as weights) and to record the number of birds killed by fishing lines. The results were startling. Over a nine-month period they found an average of 270 metres of line, 86 pieces of lead shot and five hooks per kilometre of bank. Forty-two birds of seventeen species were found killed by fishing lines. Now the catastrophic decline in the stocks of mute swans on the Thames (pp. 96–7) has been attributed to lead poisoning, caused by the consumption of lead shot. Non-toxic alternatives to lead shot are being sought, but the main solution is in the education of anglers.

Other, more extensive problems of pollution by agricultural, domestic and industrial sources are indicated on pages 84–5. These forms of pollution are often not obvious at the surface, but may act on the organisms at the base of the food pyramid, so eliminating higher consumer populations.

A CHANGING HABITAT

Change is a part of the natural patterns of landscapes and their communities, but man-made alterations of wetland habitats have had far-reaching effects on their wildlife.

The lush vegetation of a summer pond may disappear in winter, leaving a muddy margin. The thick ooze of an estuary may in time become a salt marsh. The first event is part of a regular seasonal cycle; the second is a part of long-term changes in the environment, in which each community alters the habitat to suit the next. In some places, as in peat, we can trace the evidence for these changes back for centuries, and often for millennia (pp. 52–3). The flora and fauna of these communities have often come down to us through many millions of years of changing conditions. During this time some species, such as the forest horsetails, have disappeared; others survive, such as the ancient waterlily family that is found in still waters from the tropics to temperate regions. Recent changes in climate have altered the distribution of many wetland plants, such as the cranberry, which although common in post-glacial Europe now survives in southern localities only in upland sphagnum bogs.

The greatest agent for change in the last 5,000 years has been man himself. Today there are few communities of wetland plants and animals that are not affected by man's activities, which are now threatening to destroy much of our wildlife. Man's expanding population needs food, space and fuel in ever increasing quantities. In each region of the world, food is supplied by relatively few plant and animal species; so man must provide the right habitat and nutrients for these if he is to have a good crop. Lowland wetlands often have an abundance of nutrients, so that, if the water is removed, excellent farmland is acquired. However, for many species of wild plants and animals, this will be accompanied by the destruction of their habitat.

Land from the sea
Where rivers drop their nutrient-rich load in shallow estuarine areas, man has frequently attempted to reclaim and drain this land in order to use it as farmland or settlement space. But these areas are also feeding grounds for enormous numbers of migrating and

▲The settlement of Veere in the Netherlands is on an arm of the Eastern Scheldt that has been reclaimed and drained, bringing into cultivation extensive areas of estuarine deposits that are below sea level.

▶Bewick swans *Cygnus bewickii* gathering on part of the flooded Ouse washes in winter at sunset. This important wetland site in East Anglia, created in the seventeenth century as a flood relief measure, now supports up to 60,000 overwintering wildfowl.

resident wildfowl, and large-scale alterations to the habitat may endanger these species. The environs of the North Sea estuaries have been reclaimed for centuries, but present plans to alter the tidal regime would further reduce the habitats for wildlife to a critical point.

Maintaining a balance
In the Netherlands, almost half of the cultivated land surface is below sea level, so the water must be pumped into the sea to maintain drainage of the area. Once the water is removed the land shrinks, lowering the level and thereby creating a considerable flood risk, especially from the sea. Despite the erection of sea walls and barriers, during the last flood in 1953 over 2,000 square km of land was under water and 1,843 people lost their lives.

Since this time further plans have been made to protect this coastline from the sea, and some of them implemented. But any barrier that stops the sea from flowing in also restricts the movement of sediment and alters the estuarine habitat, eliminating many marine species. Although it was originally planned to isolate the whole of the eastern Scheldt, fishermen and conservationists became aware that this would endanger the whole estuarine ecosystem. The protest movement caused the Dutch government to alter its plan, so that it is now intended to build a storm surge barrier which, although restricting tidal movements, would not prevent them. Perhaps one of the biggest problems over large-scale projects of this kind is to predict the effects on wildlife. Too often in the past we have found out only after the event.

◀Peat diggings on the Somerset Levels in south-west England, along the edge of Shapwick Heath Nature Reserve. In order to remove the peat deposit down to the marine clay, the water is pumped out. The water level in the reserve also falls, changing the conditions within the reserve.

▼The freshwater areas at Hoveton in the Norfolk Broads in East Anglia are all man-made. They consist of straightened channels with little fringing vegetation (foreground), and flooded medieval peat cuttings (background) with abundant fringing and submerged vegetation.

A new regime

There are many freshwater habitats that have been created by man for his own purposes, but have been colonized by a variety of aquatic plants and animals. Reservoirs (pp. 60–1), canals (pp. 98–9) and gravel pits (pp. 178–9) have all provided conditions where aquatic species may thrive. In other places, wetlands are produced as a side-effect of man's drainage or other activities.

The Ouse washes in East Anglia are now a site of international importance for wildfowl, but were created in the seventeenth century as a flood relief measure. They occur between the old and new Bedford rivers, and the latter, which receives water from the Great Ouse river, is tidal. In times of flooding, which occurs regularly in winter, the sluices are opened on to the washes, and the flooded meadows become the feeding or resting grounds for thousands of overwintering or migrating birds, such as ducks, geese and swans. Bewick swans rest on the washes and feed on the surrounding farmland, while the mute swans feed and breed on the washes. In the summer the land produces rich hay meadows, as well as a breeding place for birds such as the black-tailed godwit (pp. 56–7) and the ruff. If there is a summer flood these birds may suffer great losses.

This type of wetland management was formerly quite common, and hay meadows were regularly flooded with nutrient-rich water to produce a good, if late, hay crop, so they also provided a habitat for many wetland species of plants and birds. But in the battle to increase productivity many hectares of wetlands are drained and converted to arable fields every year, involving expensive flood prevention and drainage schemes. Unfortunately, drainage schemes cannot be selective. Local wetlands that were dependent on a high water level dry out and their native species disappear, unless the water level can be artificially maintained at a high level.

Peat for fuel or gardens

The peat deposits of the world cover around 230 million hectares, containing around 330,000 million tonnes of organic matter, which is a potential source of fuel, or of nutrients for growing plants. Although peat has been used as a source of fuel for centuries in most European countries where it occurs, it is at present being extracted on a larger scale than ever before. The Irish Peat Development Authority controls 50,000 hectares of peat which supply power stations as well as gardeners. In the past, peat cutting was done on a cyclical basis, the old peat cuttings flooding and providing a habitat for many wetland species to colonize. It is now thought that many of the freshwater areas of the Norfolk Broads were formed in abandoned peat cuttings of a medieval date. But on the land between, increasing numbers of drainage schemes, accompanied by drying-out of the peat, have brought about shrinkage of the land at the rate of around 2 cm a year, so the main drains are now banked up well above land level. Today, wholesale extraction of peat is facilitated by removal of the water with pumps, so that the peat-cutting machines can remove the mire habitat along with the peat. But what happens to these areas after the peat has been removed? As peat forms in a basin, most of these areas will become flooded unless the water is pumped out, as has happened in the Norfolk Broads. In Somerset, in the south-west of England, it is planned to make inland lakes after the extraction of the peat, so that some aquatic species may colonize the area, although the bog species cannot survive.

What one generation creates, another destroys, and aquatic and wetland life disappears or survives in the new regimes. The niches for wetland wildlife become increasingly confined to reserves, which cannot easily be separated from the management of the land around them. Even if we can maintain the water level or control the organic pollution, these reserves are frequently isolated from one another by a man-made 'desert' of field, forest or settlement. This creates problems of dispersal for individual species, and of food supplies for the top carnivores that require a large territory over which to hunt.

177

CONSERVATION AND MANAGEMENT

The experience gained from the successful management of wetlands and coastal sites provides invaluable hindsight for setting up new reserve areas.

Wetland sites, which are subject to heavy pressure, have to be actively managed so as to maintain a variety of species and still allow the fullest use of the water. For example, anglers prefer weed-free conditions with no overhanging branches which will snag their lines; walkers want paths, scenic views, car parks and picnicking facilities; while water birds require cover, freedom from disturbance and an abundant food supply.

Recreational disturbance by water-skiing and power-boat enthusiasts can cause major problems. A simple solution would be to prohibit access. However, such preservation will not conserve wetlands since, if they are left alone, plants gradually invade the open water and a natural succession ultimately leads to the development of woodland (pp. 48–9). It is possible to integrate recreational use with sound management of a wetland reserve.

Successful management is dependent on a thorough knowledge of the behaviour and ecological requirements of both the inhabitants and the human exploiters, as well as the understanding of how the habitat itself will respond to its utilization. A well managed wetland area can be very rich in both plant and animal species.

Managing a gravel pit

The Sevenoaks Experimental Wildfowl Reserve in Kent, in south-east England, is a reserve which has been created from old gravel pits with the full co-operation of the owner, anglers, wildfowling clubs and naturalists' organizations. Duck have territorial requirements for nesting and for places in which to 'loaf' about safely. Cross-shaped islands provide shelter whichever way the wind blows, and need to have gently sloping shorelines, so that the duck can walk ashore. Loafing areas need to be bare; whereas nesting areas need to be densely vegetated.

At Sevenoaks, some natural islands were left unexcavated; in shallow lakes others were created, while, in the deep pits, floating islands were made. A total of over 19,000 trees, shrubs and water plants were planted between 1960 and

◄A white stork *Ciona ciona* flying in to its nest in a eucalpytus tree in the Algarve, Portugal.

◄Flamingos *Phoenicopterus ruber* flying back to their breeding ground late in the day in the Camargue.

1974 to provide cover, nesting sites and food. Plantings were determined by identifying the seeds inside the stomachs of duck which had been shot in the vicinity. About 90 per cent of the Sevenoaks shoreline is now vegetated with emergent plants and overhead trees.

To increase the stock populations by attracting wild birds, ducklings and goslings of several species were reared and released after ringing. The recoveries of the rings demonstrated the effectiveness of the management programme. Various ponds were treated with a slurry of farm animal manure to increase the insect populations and thereby attract waders. New ponds were 'dug' with small explosive charges.

But management costs money, so facilities had to be created to enable visitors to see the wildfowl without disturbing them. Walkways were built and permanent hides erected. Fishing points were made, concealed by growths of marginal plants, near 'swims', or areas where fish tend to go. Fishing is not allowed on certain lakes, however, during the birds' breeding season, and it is voluntarily ceased in hard weather when the wildfowl are concentrated in a few open pools. A certain amount of pest control is needed, but predators do perform a useful function in removing diseased or weakened animals.

The Sevenoaks Reserve shows how,

◄The view from a hide showing greylag geese *Anser anser* in a sheltered backwater of Sevenoaks Experimental Wildfowl Reserve with Sevenoaks town behind.

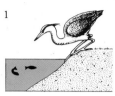

1

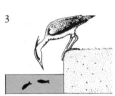

2

3

▼A duck decoy, showing the four curving, covered 'pipes' down which the ducks are lured by a specially trained dog.

▲Herons can easily reach fish in ponds with gradually sloping banks (1), but they have difficulty if a barrier is erected (2) or if the sides are vertical (3).

with imagination and a great deal of hard work, a very attractive wetland reserve carrying a diversity of species can be created adjacent to a large town. This site has been designated a Site of Special Scientific Interest (SSSI) by the Nature Conservancy Council.

Success at Minsmere

In 1947 avocets bred at Minsmere, on the east coast, for the first time in Britain for a hundred years, and in the following year it became a reserve managed by the Royal Society for the Protection of Birds (RSPB). The diversity of habitats (marsh, lagoon, reed-beds, heath and woodland) within this 607-hectare reserve and its close proximity to mainland Europe make it one of the best places in Britain for seeing a variety of birds in one day.

Inside the sea wall, the lagoon was excavated by machine and islands were created in an area known as the Scrape. Here, avocets and terns breeding in summer, and migrant waders gathering in spring and autumn, can be seen from the public hides. Optimum conditions are provided for the birds throughout the year by delicately balancing and finely controlling the water levels and the salinities using sluice gates on both the seaward and the landward side. The terns were encouraged to breed on the new islands, after gravel had been collected and ferried in sandbags by boat to the islands. The Scrape has been enlarged annually, which has encouraged more birds to come in to breed, until, in 1980, 58 pairs of avocets were breeding at Minsmere.

Fluctuating populations

The Camargue is a triangular area on the south coast of France which is bordered by the two arms of the Rhône and by the sea. The chief attraction of this region for naturalists is the flamingos, which breed on flat islands in the salt pans (pp. 102–3). These islands became eroded, so the area suitable for breeding dwindled. The construction of an artificial island as well as some artificial mud nest mounds (made by packing mud into buckets) attracted flamingos to breed in large numbers, so that now some 20,000 birds spend the summer in the Camargue.

The white storks have not fared so well. The draining of fens and the use of pesticides have contributed to their decline in Europe since the end of the nineteenth century. Between 1938 and 1954 the European population was halved and the stork is now extinct in parts of west and north Europe, but it is still quite common on open farms in east Europe, where it feeds on frogs, snakes, fish and small mammals. Storks take advantage of man-made steeples and telegraph poles as high-rise nesting sites, as well as trees. Active measures to slow down the decline of the stork population include the building and renovation of their nests and the removal of frogs' legs from the menus of some restaurants.

Duck decoys

Duck decoys, used for catching wild ducks for food, originated in Holland in the late sixteenth century. At the edge of a lake, or around a purpose-built pond, curved, tapering ditches, known as 'pipes', are built. Each pipe is covered over by a netting stretched over a series of semi-circular hoops, so that it becomes a tunnel. A specially trained dog lures ducks into the open end of the tunnel by appearing and then disappearing behind screens as it runs alongside the tunnel up to the end. When plenty of ducks have entered the pipe, a decoyman appears at the opening, which drives the birds down to the end, where they are caught in a net. Today, only a few decoys remain in Britain, where they are used for catching and ringing birds; but they are still used in Holland for catching ducks to eat.

Unwelcome visitors

Reservoirs attract birds not only for feeding, but also for nesting and roosting. In the Strathclyde region of central Scotland, a pollution problem was caused by the droppings of some 5,000 gulls using two reservoirs as their night-time roosts. Covering the water surface would have been prohibitively expensive, but the problem was solved by using a loudspeaker to play the gulls tape-recordings of their distress calls two hours before dusk, and so persuading them to roost elsewhere.

A detailed study has shown that predation by herons on fish farms is widespread. The daily intake for an adult heron is equal to two almost fully grown trout or twelve fingerlings.

As a fully protected bird in parts of Europe, the heron cannot be legally shot. The predation by herons can be prevented expensively by caging the fish farm or by using scaring devices, which are effective only in the short-term. There are other deterrents which are both relatively inexpensive and effective. Herons normally feed from the bank, by bending down to the water. If the banks are steepened, or the water level lowered, they cannot reach the water. Alternatively, a 35cm-high barrier can be erected around the edge of the pond.

Mismanaging the sea

Conservation of renewable resources in the sea, such as stocks of fish and whales, has always presented serious problems (pp. 168–9). The industries needed substantial financial investment in ships and equipment; once spent, there was the demand to maximize the return on the investment. Consequently, whales were hunted down to unacceptably low levels, despite the good intentions of the International Whaling Commission to limit catches. Similarly, attempts to manage stocks of commercial fish such as herring, plaice and cod have been frustrated, partly by economic pressures and partly by the lack of understanding of the size of natural fluctuations in the stocks produced by long-term climatic changes. Herring were finally reduced to uneconomic stock levels by the use of industrial fishing techniques to catch vast quantities of juvenile fish to make fish meal for feeding to stock.

PHOTOGRAPHY

Water – in both liquid and solid forms – provides endless opportunities for the pictorial photographer, while the plants and animals which live in or are associated with water will appeal more to the naturalist photographer.

The motivation for taking photographs will vary, but the aim should always be to spend some time in selecting the best viewpoint and camera angle.

Landscapes and habitats

Any type of camera can be used to photograph wetlands and coastlands, but a single lens reflex (SLR) 35mm camera with an interchangeable lens will be more versatile than a camera with a fixed lens. The standard lens for such a camera has a focal length of 50 mm and can be used for taking landscapes. Wide angle lenses which have a shorter focal length (e.g. 24 mm or 35 mm) and long focus lenses which have a longer focal length (e.g. 100 mm or more) can also be used for landscapes. Since wide angle lenses have a greater angle of view than a standard lens, they give a more panoramic picture; the narrow angled long focus lens concentrates on a detail in the landscape.

Lighting is all-important for landscape photography. Each season has its own quality of light, which also changes with the time of day. When the sun is shining early in the morning or late in the evening, it produces low-angled lighting, coupled with long shadows; whereas overhead mid-day sun gives a harsh lighting. The direction of the sun in relation to the camera also affects the mood of a photograph. Front lighting may be a safe way to get a photograph, but it is also rather an unimaginative one. Side lighting, with long shadows, is ideal for showing landscape relief. Dramatic lighting may arise when the camera is pointed towards the sun, so that it back-lights the subject, but care must be taken that the sun does not shine directly into the lens. A lens hood will help to reduce the likelihood of direct rays hitting the lens and thereby causing flare.

Bright, sunny days, however, are not necessarily best for every kind of photography. Dull, overcast days provide a soft, gentle lighting which is ideal for showing petal detail of white flowers. Interesting pictures of moving

◄A long lens picks out an interesting design of a braided stream criss-crossing over an outwash plain in Iceland.

►A macro lens was used for this detailed close-up of shells from Shell Beach on the island of Herm in the Channel Islands.

▼Photographed against the light, looking towards the sun, the ripple pattern on an exposed sandy beach is dramatized.

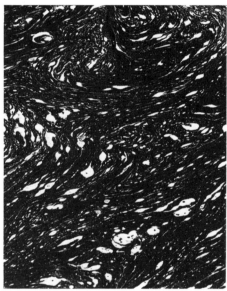

▲Turbulent river water provides endless opportunities for photography, as shown here by this abstract design of swirling surface bubbles.

►As mud dries out it shrinks, and cracks appear which join up to form an intriguing polygonal pattern. By using a macro lens the whole of the frame was filled with the centre of a dried-up salt pan.

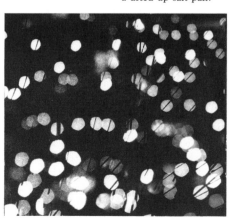

▲If sun brightlights on water are photographed out of focus, they will appear as polygons (or whatever the shape of the lens iris), which make an interesting puzzle picture. The black lines are reeds growing beside the water.

water can also be taken on dull days, by using a slow shutter speed of ½ or 1 second, which gives a soft, blurred movement. For such long exposures, it is essential to use a sturdy tripod to eliminate any chance of camera shake.

One of the main problems of photographing flat aquatic habitats from the ground is in locating a high viewpoint, so as to look down on to a meandering river or a lowland lake. A high level bridge crossing a river is ideal for a view looking up or down a river (p. 69), whereas a hill or cliff adjacent to a river or the coast is often worth climbing for a better view (p. 154).

When using a rectangular 35mm format, you can choose whether to take a horizontal or a vertical picture. Many more 35mm photographs are taken as horizontals, simply because this is the most convenient way of holding the camera. Yet by turning the camera through 90°, quite a different feeling to the picture can be gained. Many vertical photographs have been taken for this book, as well as horizontal and square (on a 6 × 6 cm format) ones; but the shapes which finally appear on the printed page are the ones created by the designer, who in some cases crops the original image.

Close-ups

Throughout this book, there are many examples of close-up photographs of small animals and plants. Such pictures, providing they are sharp, give much more information about the shape and structure of an individual species than can be seen in a general location picture. The easiest (and cheapest) way to take close-ups is simply to attach a close-up lens to the front of the standard lens, including fixed-lens cameras. Greater magnifications can be gained with an SLR camera by removing the lens and inserting extension tubes or bellows between it and the camera body. For anyone who intends to take a lot of close-ups, a special macro lens, with its own built-in extension, is ideal, since the focusing range is continuous from infinity up to

half life size. The main disadvantage of extension tubes, bellows and macro lenses, however, is that since they all move the camera lens further away from the film plane they reduce the light reaching the film, thus making an exposure increase necessary.

Critical focusing is of greater importance when taking close-up photographs than when taking landscapes, since the depth of field (the zone of sharp focus) is more limited. Although it is quite possible to take close-ups with a hand-held camera, better quality pictures will be achieved if the camera is used on a tripod, since the depth of field is a function of the image size and the size of the lens aperture. Therefore, it can be increased either by moving further away to get a smaller image, or by using a slower shutter speed and stopping down the lens to a smaller aperture (e.g. from f/5.6 to f/11). When taking a close-up of a static plant, such as lichen growing on a rock, it is possible to use a very slow shutter speed (½ second) coupled with an aperture of f/16, so as to get everything appearing sharply in focus.

Long lenses

These lenses are often mistakenly referred to as telephoto lenses, but by no means all long lenses are of a telephoto design. Although long lenses can be used for taking landscapes, they are most often used for taking birds and mammals in the wild. Great satisfaction is undoubtedly achieved by stalking a subject in order to get it almost filling the frame, but such a picture gives no information about where the bird lives, feeds or breeds in its natural habitat. As when photographing flowers, pictures which show something of the surrounding terrain or vegetation can be much more informative, and perhaps ideally should be used in conjunction with impact portraits. (Examples of bird photographs which also illustrate their habitat can be seen on pp. 46 and 160–1).

As with close-ups, great care must be taken when focusing a long lens, which should also be used supported either on a tripod, or on a shoulder pod, which can be braced against the body as the camera is panned round to follow a bird in flight. An ideal place to photograph flying birds is from a cliff top, directly above where seabirds are nesting. The birds will then fly past as they take off and return to their nest sites. If there is a strong head wind, the birds may gain little headway and so remain almost stationary for a while. To avoid blurred wings of flying birds, a fast shutter speed of $1/250$ or even $1/500$ second should be used.

Good behavioural photographs are rarely achieved quickly. Time spent observing a bird or mammal will mean

that its behaviour is more likely to be predicted, which will save film in the long run!

Photographing through water

One of the delights of giving rivers, ponds or rock pools more than a passing glance is in seeing the infinite variety of the colours and shapes of the life below the surface. Anyone who has looked down into water wearing Polaroid sunglasses will know how effective they are for cutting out the surface skylight reflections so that the plants and fish become clearly visible.

One solution for getting clear photographs through water is to use a polarizing filter on the front of the lens. By rotating the filter and looking through an SLR camera, a grey water surface will gradually clear and reveal the life below. For maximum effect, a polarizing filter should be used by holding the camera at an angle of 37° to the water and not directly overhead. The main disadvantage of using this filter is that it reduces the amount of light reaching the film by as much as 1½ stops, which may make it difficult to photograph moving subjects with a slow colour film.

A much cheaper way of cutting out the sky reflections on a small area of water is simply to hold an umbrella (or a dark mackintosh) over it.

Patterns and designs

The ways in which water moves or ice crystallizes make exciting patterns and designs, which may also exist in rock or sand sculptured by water. Many of these images with strong shapes make ideal subjects for black and white prints.

Aquarium photography

All the techniques so far described apply to outdoor photography, but the only way of showing the structure of small aquatic organisms – especially those which live in murky waters – is to photograph them in aquaria. Only a few animals should be collected for photography and care must be taken not to let their container be warmed up (for instance, by the car heater on your journey home). Animals which live in well oxygenated mountain streams are particularly susceptible to death from lack of oxygen.

Since Perspex easily scratches, glass aquaria are preferable for photography. Even then, the front glass must be absolutely clean. A matt black mask attached to the front of the camera lens (with a central hole cut out) will eliminate distracting reflections of the camera in the aquarium glass.

Small electronic flashlights are ideal for aquarium photography, since they do not warm up the water, and they will stop the movement of most animals.

181

WHERE TO GO

This book shows that there is life in every sort of water body. Do observe as many different kinds of habitat as you can. If you have easy access to a stream or pond, go there often; you will be surprised at the seasonal changes and by the occasional visitors from other habitats.

When you have become familiar with the water bodies around you, it will be worth contacting national bodies which can provide you with information on other wetland and coastal areas. Regional natural history societies or conservation trusts will be able to advise on their particular areas. Join one of these and you will be able to visit interesting places where access is limited. For those wishing to study a subject in more detail, organizations such as the Field Studies Council run residential courses in Britain and on the European mainland.

If you are travelling to a new area, you will find detailed maps, identification books containing distribution maps, and reliable local guides invaluable. A map showing as many natural features as possible will help you to predict environments. Contour lines will show highland and lowland areas, as well as indicating types of rivers and lakes; closely spaced contours are associated with swift mountain streams, and absence of contours with meandering rivers that may or may not be eutrophic, according to rock type and land management. A geological map will help you to predict the type of water that you can expect to find. Depending on its scale, your map may give you the nature of coastal deposits of sand, mud or rocks, as well as inland vegetation types, such as moorland, wetland and forest.

Many coastal and estuarine areas have been declared reserves because they are important sites for overwintering waders or wildfowl, or for breeding seabirds; inland wetland sites are often important for their communities of plants and animals rather than for a single species. The map below contains a selection of both types of site, and some areas of beauty that have a variety of habitats within them.

Localities

Iceland
1. North-west peninsula: area of Tertiary basalt with fjord-like coastline; high cliffs; breeding seabirds.
2. Snaefellsness peninsula: sea cliffs; inland pools, lagoons, glacier; seabirds.
3. Hunafloi: rivers, lagoons, salt marsh; rocky coast around Blonduos.
4. Mývatn: lake, islands, pools, river; breeding birds.
5. Thingvallavatn: lake; breeding birds.
6. Westmann islands: rare seabirds.
7. Skaftafell and coast to the east: national park, glacial scenery; moorland, freshwater and coastal species.

Norway
8. Hornvika island reserve (P): breeding seabirds.
9. Varangerfjord: only mainland coastal haunt of Arctic seabirds in Europe.
10. Vesterbotten, Stabbursneset: coastal marshes; migrating birds, moulting eider duck.
11. Lofoten islands: coastal lowlands, sandy beaches; seabirds.
12. Forra valley: mires, bogs, small lakes; breeding birds including divers, ruffs.
13. Havmyran, Hitra island: mires, bogs, small lakes; abundance of birds.
14. Smøla archipelago: coastal peatlands, islands; nesting, moulting and migrant water fowl.
15. Djupvatn: lake; at northern end, extensive marshes, bogs, fell country.
16. Dovrefjell: Fokstumyra national reserve; marshland, lake, fell.
17. Atnasjøen: lake of Atna river; erosion and deposition patterns; aquatic species.

18. Rønnasmyra: raised bog complex.
19. Øra: eutrophic, brackish water shallows in Glommer delta; flora and fauna.

Finland
20. Åland islands: on edge of the tideless Baltic Sea; vegetation; seabirds, birds of prey including white-tailed eagle.
21. Inari: largest lake in Finland; marshes, islands, inlets; breeding birds.
22. Oulanka National Park: dramatic scenery of river Oulankajoki.
23. Linnansaari National Park: islands on Lake Sainau; elk, common seal, osprey.
24. Aspskär: seabird reserve island off Lovisa; only Finnish haunt of Caspian terns.

Great Britain
25. Shetland islands: cliffs; seabirds.
26. St Kilda: islands, stacks, spectacular cliffs; nesting seabirds; largest gannetry in Europe.
27. Farne islands: coastal vegetation; birds.
28. Caerlaverock: extensive coastal marshes on Solway Firth; overwintering geese and waders.
29. Lake District: dramatic upland scenery; lakes, rivers; moorland species.
30. Snowdonia: mountains, moors, bogs, lakes.
31. Ynyslas: sand dune reserve, slacks, raised bog nearby; interesting dyke flora.
32. Pembrokeshire coast and islands: sand dunes, salt marshes, cliffs, islands; nesting seabirds.
33. Wye valley: river system; much unspoilt countryside.
34. Blakeney Point: wet meadows, reed beds, salt marshes; breeding and visiting birds.
35. Norfolk Broads: wetlands with reed beds and open water.
36. Ouse washes: meadows subject to winter flooding, attracting wildfowl, especially Bewick's swans.
37. Minsmere (P): RSPB reserve; salt marsh, reed beds, open pools.
38. Havergate island: RSPB reserve in Ore estuary; breeding waders including avocets; vegetated shingle.
39. Dungeness: shingle promontory; landfall for spring and autumn bird migrants; bird observatory; most extensive area of vegetated shingle in Europe.
40. The Lizard: coastal area; plants; birds.
41. Somerset Levels: ditches, dykes, wet meadows, peatlands; reserves.
42. Slimbridge, Severn estuary: largest collection of wildfowl in world; visiting ducks, geese, swans.
43. New Forest: pools, streams.

Ireland, Northern
44. Lough Foyle: coastal shallow lough; estuarine and bird life.
45. Lough Neath, Lough Beg: shallow, reed-fringed loughs; wintering wildfowl.
46. Strangford Lough: tidal mud flats; birds of passage.

Ireland, Republic of
47. Lough Suilly: sea lough, salt marsh, wet meadows; aquatic vegetation, birds.
48. Lough Corrib: freshwater loughs on carboniferous limestone.
49. Shannon river: river valley, extensive mires, wet meadows, frequently flooded in winter, estuary, reedswamp, salt marsh.
50. Akeragh Lough: brackish lagoons, southern one rich in vegetation; wildfowl.
51. Wexford slobs: reclaimed marshland; salt marsh, sand dunes; flora in ditches; overwintering wildfowl.
52. Dublin Bay, including North Bull island and Lambay: tidal mud flats, salt marsh.
53. Lough Owel, Lough Durravaragh: shallow, clear-water, limestone loughs.
54. Lough Oughter: maze of islands and peninsulas created by drowned drumlins; reed beds, alder swamp.

Sweden
55. Abisko National Park, edge of Lake Torneträsk: lake, marshes, meres; Arctic species of birds, plants.

56. (a) Muddus; (b) Sjofallets; (c) Padjelanta: national parks; mountains, mires, lakes.
57. Peljekaise National Park: mountain valleys, lakes, marshes, meadows; wildfowl.
58. Ånnsjön: vast, shallow lake, lagoons, mires, marshes; waders, other birds.
59. Gammelstadsviken: shallow bay; northern flora and fauna.
60. Hjälstaviken: bay on Lake Malaren; reedswamp, open water, wet meadows; bird reserve.
61. Kvismaren: wetland, swampy meadows where two lakes have been drained for waders, wildfowl.
62. Tåkern: extensive reed beds around lake, submerged vegetation; wildfowl.
63. Kävsjön, Store Mosse: raised bog and mire complex, open water.
64. Oland: island in Baltic Sea; coast at Södviken and Ottenby good for migrant and breeding birds.
65. Gotland and islands off east coast: migrant birds.
66. Falsterbo: peninsula of shallow coastal waters, sandbanks; inland lagoons, mires.

Denmark
67. Skagen, Hirsholmene: vast dune systems; birds.
68. Limfjorden, Vejlerne, Tyboron: wetlands, coastal marshes, lagoons.
69. Nisum Fjord to Stadilfjord: shallow, brackish water lake, reed beds, coastal marshes; spring and autumn migrant birds.
70. Ringkøbing: brackish water fjord, reed beds, coastal marshes; migrant birds.
71. Vorsø: islands; rocky coast in Horsens Fjord.
72. Fyn, Langeland: lakes, salt flats, islands.
73. West Sjaelland: includes Hovestrand, sandy bay, reed beds; Reerso saltings; and Tisso, large, shallow lake, multitudes of waders.
74. Lolland, Falster, Møn: islands; reserve at Nakskov Indrefjord; Ulfshale, area of marshes, birds.

Germany, West
75. Wattenmeer, Wadden Sea: mud flats, salt marsh, sand dunes (continued to Husum, Denmark).
76. Heligoland: island reserve; migrant birds, breeding guillemots, kittiwakes.
77. Niederelbe: sand banks, islands, mud flats, reed fringe, dykes, riverine marshes.
78. Kiel Bay, on Baltic Sea: dunes, dykes.
79. Lakes in Schleswig-Holstein: glacial lakes; vegetation; birds.
80. Dümmer reserve: large lake; reeds, submerged species; wet meadows; alder carr.
81. Heiliges Meer: sandy soil, acid pools, bogs.
82. Wetlands along Rhine on German and French banks.
83. Bodensee: once sub-Alpine oligotrophic lake, now (through man's activities) mesotrophic lake; peat bogs, damp meadows.
84. Federsee: reed marshes; peatland vegetation; birds.
85. Unterinn: four reservoirs; reedy edges, islands; waterfowl.

Germany, East
86. Rostock area and Baltic coast, particularly Darss reserve: breeding area of crane.
87. Müritz See: largest inland lake in East Germany; eutrophic lake; plants; wildfowl.
88. Gordsdorf, Pinnowsee: wetland reserves for beaver.

Netherlands
89. Wadden Sea: mud flats, salt marshes inland from Texel; plants; birds; seals.
90. Friesian islands, including Texel: dunes, marshy slacks, lakes, dykes, inland marshes; many species.
91. Zwartemeer: eastern extension of Ijsselmeer; eutrophic, shallow lakes, reed beds, breeding birds.
92. Naardermeer: freshwater fens, marshes, swamps.
93. Biesbos: until 1970 (when dam was built) a tidal area. Network of marshland, swampy woodland; breeding birds.

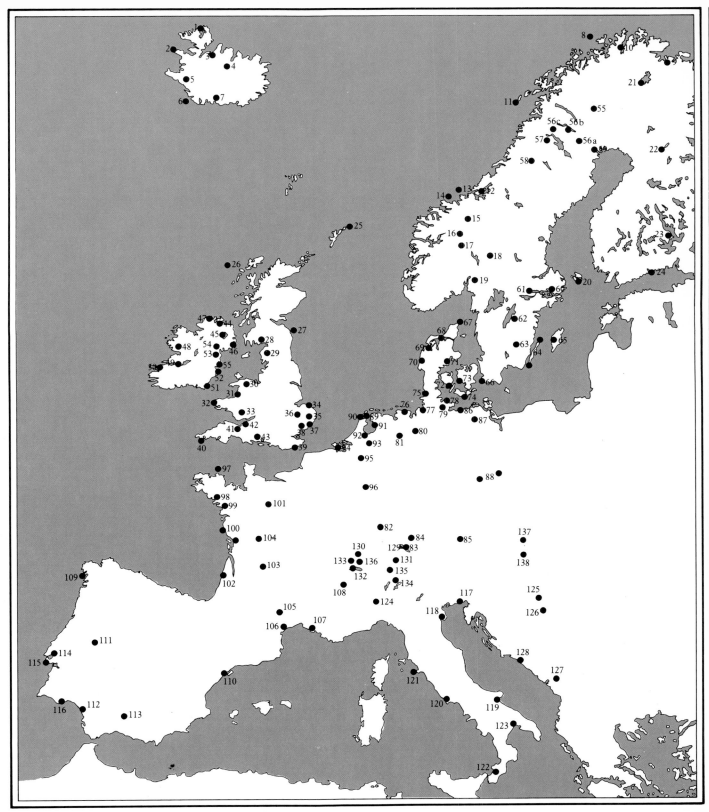

94. Rhine delta: includes Oosterscheldt, sheltered inlet of North Sea, saltings, shallows, mud flats; and Quackjeswater, nature reserve of mesotrophic dune lakes; rare plants; birds.

Belgium

95. Hautes Fagnes Nature Reserve: extensive area of peat moss.
96. Blankaart: lake, extensive reed beds, permanent water meadows.

France

97. Sept Îles: cliff islands; reserve for breeding seabirds.
98. Golfe de Morbihan: freshwater lagoons, salt marsh area around Vannes; plants; birds; wetland reserve north of Nantes.
99. Lac de Grande Lieu: open water reedswamp; birds.
100. Île de l'Olonne: extensive tidal salt marsh area, brackish water lagoons; inland reedy pools, fish ponds.
101. Sologne: damp heath, ponds, floating plants, reedswamp fringes.
102. Basin d'Arcachon: fresh and salt-water marshes adjacent, tidal creeks, inland ponds. Extensive sand dunes to south.
103. Dordogne: river system, limestone gorges above Souillac.
104. La Brenne: lakes, ponds, north of river Creuse; extensive reed fringes; breeding birds.
105. Gorges du Tarn: impressive limestone gorges; river.
106. Étangs de Languedoc: saline and freshwater pools, lagoons.
107. Camargue: Rhône delta; salt marshes, brackish and freshwater marshes, dunes; breeding birds include flamingos.
108. Lac du Bourget: lake, extensive reed beds, marshes; birds.

Spain

109. Rias Bajas: fjord-like bays between La Coruña and Vigo; overwintering ducks feed on eel grass.

183

110. Ebro delta: extensive rice paddies; marshland, saline or brackish water lagoons.
111. Tablas de Daimiel: extensive marshlands created by rivers in the Guadiana; fresh and brackish water lakes, ponds; birds; reservoirs.
112. Coto Doñana: national park; one of largest coastal ecosystems in Europe; dunes, salt marsh, variety of habitats; rare breeding birds.
113. Laguna de Fuente de Piedra: salt lake; flamingo colony, other birds.

Portugal
114. Tagus estuary: complex of mud flats, salt marsh, shallow lagoons, polders, dunes, damp slacks.
115. Sesimbra, Lagoa de Albufeira: variety of habitats.
116. Faro-Tavira, on Algarve coast: maze of islands, peninsulas; extensive sand bar.

Italy
117. Lagune di Marano: large brackish to salt-water lagoons; submerged plants; birds.
118. Lagoons between river mouths of Po and Adige: used for fish farming but relatively undisturbed. Also estuarine mud flats, swamp woodland.
119. Gargano peninsula: salt pans, lagoons; autumn migrants.
120. Laghi Pontini: part of sand-dune-locked-lake system with variable salinity.
121. Lago di Burano: similar to above; variety of plant species.
122. Messima river valley: rich in species, especially in limestone area.
123. Lago di San Giuliamo, near Matera: water courses; plants; autumn migrants.
124. Upper valley of Po, around Vercelli: ancient irrigation channels to rice paddies; flora.

Yugoslavia
125. Baranja: Danube–Drava confluence; marshes, peat bogs, flood plains, ponds, reed beds; reserve of Kopacki rit (P).
126. Obedska bara: large area of marshes, swamps, open water; small area is restricted reserve.
127. Skadarsko Jezero: vast lake on Albanian border; extensive reed beds, maze of lagoons, wet meadows regularly flooded.
128. Metković, on lower Neretva river: marshland area around reserves at Hutovo Blato and Deransko Jezero; lakes, ponds, drainage ditches.

Switzerland
129. Bodensee: see under Germany, West.
130. Lac de Neuchatel (P): open shallows, marsh zone; waders.
131. Lauerzersee: lake; marsh area at north-eastern end of Lake Lucerne.
132. Lac Leman (Lake Geneva): important site for migrating and overwintering wildfowl.
133. Lac Brenet: oligotrophic mountain lake; marsh area.
134. Lago di Muzzano: moraine-formed lake.
135. Lago di Ritom: lake with sulphur bacteria.
136. Chavornay clay pits: area of reeds, willow scrub; breeding birds.

Austria
137. Marchauen, Marchegg: wet meadowland along river March on border of Czechoslovakia.
138. Neusiedlersee: large, eutrophic lake; extensive reed beds; many rare birds. To the east, Seewinkel: mosaic of eutrophic ponds.

(P) permit required.

USA
National Parks in USA which are Coastal (●) or which contain large amounts of fresh water (△)

△ *North Cascades National Park.* 1600 square km of virtually untouched landscape, the northern part of which borders on Canada. Breathtaking alpine scenery with glaciers and lakes.
△ *Mount Rainier National Park.* Glaciers, glacial rivers and lakes. Mount Rainier is highest peak in Cascade Range.
△ *Crater Lake National Park.* Lies in Cascade Range. Lake is deepest in US (600 metres).
△ *Lassen Volcanic National Park.* Scattered all over the Park are fumaroles, hot springs, boiling mud pots, and vents; also Manzanita Lake.
△ *Yosemite National Park.* Glaciers, rivers, waterfalls.
△ *Zion National Park.* Falls and gorges cut by the Colorado River and other rivers.
△ *Rocky Mountain National Park.* Glacial lakes, waterfalls and streams.
△ *Grand Teton National Park.* Rivers, lakes and marshlands, waterfalls.
△ *Yellowstone National Park.* Geysers (including Old Faithful), fumaroles, hot springs and bubbling mud pots – most extensive thermal area on earth. Grand Canyon of Yellowstone River. Yellowstone Lake.
△ *Glacier National Park.* Alpine-glaciers, hundreds of lakes, waterfalls.
●△ *Voyageurs National Park.* Network of lakes and streams. Numerous birds/waterfowl feed in the bays, lagoons and kettleholes. 3 big lakes, small lakes, gorge, and waterfall. Bogs, cliffs and sand beaches at water's edge.
△ *Isle Royale National Park.* Main island in Lake Superior surrounded by 200 small islands. Streams connect with inland lakes, swamps. Fjord-like rocky coast with deep bays.
●△ *Acadia National Park.* Finest surviving fragment of New England shoreline. Islands, peninsula. Coves and harbours bordering Atlantic Ocean. Glacial lakes and low wetlands.
△ *Hot Springs National Park.* Hot springs in which rare species of blue-green alga thrives.
●△ *Everglades National Park.* Swamps, much of which diked, drained and developed in past 30 years. Brackish water, mangroves. Low islands (keys) in Florida Bay.
Other areas of interest:
● *Big Sur, Point Lobos on Californian Coast.* Spectacular coastal scenery, seabirds and sea otters.
△ *Okefenokee, Georgia.* Swamp area with swamp cypresses and many insectivorous plants.
●△ *The Bayous of Louisiana.* Water from the Red and Mississippi Rivers feed a system of freshwater swamps and salt marshes.

Throughout the United States there are numerous State Parks, Wildlife Refuges and Wildlife Management Areas, details of which can be obtained from each state.

SOCIETIES TO JOIN

General Interest
Association for the Preservation of Rural Scotland
20 Falkland Avenue, Newton Mearns, Renfrewshire G77 5DR, Scotland
British Naturalists' Association
43 Warnford Road, Tilehurst, Reading, Berkshire, England
British Trust for Conservation Volunteers (BTCV)
36 St Mary's Street, Wallingford, Oxfordshire OX10 0EU, England
Runs a voluntary field force that carries out practical conservation chores.
Committee for Environmental Conservation (CoEnCo)
Zoological Gardens, Regent's Park, London NW1 4RY, England
Council for Nature
Zoological Gardens, Regent's Park, London NW1 4RY, England
Council for the Protection of Rural England (CPRE)
4 Hobart Place, London SW1W 0HY, England
Council for the Protection of Rural Wales (CPRW)
14 Broad Street, Welshpool, Powys SY21 7SD, Wales
Fauna Preservation Society (FPS)
Zoological Society of London, Regent's Park, London NW1 4RY, England
Field Studies Council (FSC)
Director & Information Office: Preston Montford, Montford Bridge, Shrewsbury SY4 1HW, England
London Office: 9 Devereux Court, Strand, London WC2R 3JR, England
Independent charity encouraging fieldwork and research out-of-doors.
Izaak Walton League of America
1800 North Kent Street, Suite 806, Arlington, VA 22209, USA
National Conservation Corps for Scotland
70 Main Street, Doune, Perthshire, Scotland
National Trust
42 Queen Anne's Gate, London SW1H 9AS, England
National Trust for Scotland
5 Charlotte Square, Edinburgh EH2 4DU, Scotland
National Wildlife Federation
1412 16th Street NW, Washington, DC 20036, USA
Royal Society for Nature Conservation (RSNC)
The Green, Nettleham, Lincoln LN2 2NR, England
Scottish Field Studies Association
Forelands, 18 Marketgate, Crail, Fife KY10 3TL, Scotland or The Warden, Kindrogan Field Centre, Enochdhu, Blairgourie, Perthshire PH10 7PG, Scotland
Scottish Wildlife Trust
8 Dublin Street, Edinburgh EH1 3PP, Scotland
Sierra Club
530 Bush Street, San Francisco, CA 94108, USA
Ulster Trust for Nature Conservation
c/o Inver Cottage, 67 Huntley Road, Banbridge, Co. Down, Eire
Watch
c/o The Royal Society for Nature Conservation, 22 The Green, Nettleham, Lincoln LN2 2NR, England
Aims to educate young people about nature and the environment.
Wildlife Youth Service of the World Wildlife Fund
Wildlife, Wallington, Surrey, England
Encourages 5–18 year-olds to play a greater part in nature conservation.
World Wildlife Fund
Panda House, 11–13 Ockford Road, Godalming,

Surrey GU7 1QU, England
Young Zoologists' Club (XYZ Club)
The London Zoo, Regent's Park, London NW1
4RY, England
Encourages children under 18 to take an interest
in animals.

More Specific Interest
Algae
Phycological Society
Hon-Secretary, c/o Department of Brewing and
Biological Sciences, Heriot-Watt University,
Chambers Street, Edinburgh EH1 1HX, Scotland

Birds
British Trust for Ornithology (BTO)
Beech Grove, Tring, Hertfordshire HP23 5NR,
England
Royal Society for the Protection of Birds
(RSPB)
The Lodge, Sandy, Bedfordshire SG19 2DL,
England
Wildfowl Trust
The New Grounds, Slimbridge, Gloucestershire
GL2 7BT, England
Young Ornithologists' Club (YOC)
The junior branch of the RSPB, The Lodge,
Sandy, Bedfordshire SG19 2DL, England

Freshwater
Freshwater Biological Association (FBA)
The Ferry House, Far Sawrey, Ambleside,
Cumbria LA22 0LP, England

Geological
Geologists' Association
Burlington House, Piccadilly, London W1V 0JU,
England

Insects
Amateur Entomologists' Society
23 Manor Way, North Harrow, Middlesex,
England
British Trust for Entomology
41 Queen's Gate, London SW7 5HU, England

Mammals
Mammal Society
Harvest House, 62 London Road, Reading,
Berkshire RH1 5AS, England
Otter Trust
Earsham, Near Bungay, Suffolk, England

Marine and estuarine
The Cousteau Society
777 Third Avenue, New York, NY 10017, USA
**Estuarine & Brackish Water Biological
Association** (EBWBA)
The Secretary, Department of Zoology,
University of Cambridge, Downing Street,
Cambridge CB2 3EJ, England
International Oceanographic Foundation
3979 Rickenbacker Causeway, Miami, FL 33149,
USA
**Marine Biological Association of the United
Kingdom** (MBA)
The Laboratory, Citadel Hill, Plymouth, Devon
PL1 2PB, England
Scottish Marine Biological Association (SMBA)
Dunstaffnage Marine Research Laboratory, P.O.
Box 3, Oban, Argyllshire, Scotland
Porcupine Society
Hon. Editor of Porcupine Newsletter, c/o The
Dove Marine Laboratory, Cullercoats, North
Shields, Northumberland, England
Concerned with promoting interest in the ecology
and distribution of marine fauna and flora in the
north-east Atlantic. Named from the 19th century
survey vessel *Porcupine*.

Molluscs
**Conchological Society of Great Britain and
Ireland**
51 Wychwood Avenue, Luton, Bedfordshire
LU2 7HT, England
Devoted to study of molluscs, including their
shells.

Malacological Society
Hon. Secretary, c/o Department of Science,
Bristol Polytechnic, Redland Road, Bristol BS6,
England
Arranges meetings and lectures relating to living
and fossil molluscs.

Plants and ferns
Botanical Society of the British Isles
c/o Department of Botany, British Museum
(Natural History), Cromwell Road, London SW7
5BD, England

ORGANIZATIONS CONCERNED WITH THE COUNTRYSIDE

**An Roinn Tailte Ant Seirbhis Foraoise & Fia-
Dhulra**
(The Department of Fisheries, Forest and
Wildlife Services) 22 Upper Merrion Street,
Dublin 2, Eire
British Waterways Board
Melbury House, Melbury Terrace, London NW1
6 JX, England
Council for National Parks
4 Hobart Place, London SW1W 0HY, England
Countryside Commission
John Dower House, Crescent Place, Cheltenham,
Gloucestershire GL50 3RA, England
Countryside Commission for Scotland
Battleby, Redgorton, Perth PH1 3EW, Scotland
Nature Conservancy Council (NCC)
19/20 Belgrave Square, London SW1X 8PY,
England
Water Space Amenity Commission
1 Queen Anne's Gate, London SW1H 9BT,
England
Co-ordinating link between recreational and
national water interests.

GLOSSARY

Abdomen area of body containing stomach and
intestines; posterior part of body
Aerobic in presence of oxygen
Aestivate dormant during drought, hot or
summer season
Alevin young trout or salmon which hatches
from egg bearing a yolk sac
Algae major group of uni- or multi-cellular,
photosynthetic plants particularly important in
the aquatic environment
Alluvium material transported and deposited by
a river
Ambient temperature temperature of
surrounding medium
Amphibia class of vertebrates with aquatic, gill-
bearing larvae and air-breathing adults which
are generally terrestrial except when they
breed; includes frogs, toads, newts and
salamanders
Amphipod member of an order of crustaceans
(marine and freshwater) which includes
sandhoppers and freshwater shrimps
Amplexus mating embrace of frogs and toads
(during which eggs are shed and fertilized)
Anadromous ascending rivers to spawn
Anaerobic in absence of oxygen
Annelid member of a phylum of segmented
worms which includes earthworms, leeches
and marine polychaetes
Anoxic deficient in oxygen
Antenna jointed, whip-like mobile sense organ
on head of various arthropods
Anterior near front (usually head) end
Archegonium female sex organ in liverworts,
mosses and ferns
Arthropod member of an invertebrate phylum
of segmented animals with exoskeleton which
includes insects, spiders, crustaceans etc.
Asexual reproduction reproduction without
fusion of gametes
Autotomy self-amputation of part of body
(particularly crabs)
Axil angle between a leaf (or branch) and stem
on which it is borne

Bacteria group of unicellular, microscopic
organisms combining both animal and plant
characteristics
Baleen horny plates attached to upper jaw of
baleen whales, used for filtering krill
Bar generally coastal, beach-like deposit
extending across a bay, inlet or estuary
Barbel tactile process on head of some fish
Bedding plane junction between layers of
sedimentary rock, representing environmental
change; these can later be deformed by earth
movements and pressures
Bedrock solid rock layer occurring beneath
loose deposits, e.g. topsoil
Biochemical oxygen demand see BOD
Biodegradable any material which can be
broken down by micro-organisms
Biological clock internal mechanism enabling
organism to 'tell the time' and carry out
certain rhythmical metabolic and behavioural
processes
Biomass total weight of organisms per unit area
or volume
Bivalve mollusc with body enclosed by two shell
valves
Bloom concentration of phytoplankton on or
near water surface
BOD Biochemical Oxygen Demand; measure of
organic pollution in water; low value=little,
high value=much
Boulder clay type of mixed glacial deposit,
broken down and transported by ice, and left
behind after the ice melted
Boundary layer narrow layer overlying the
surface of stones in a stream (0.5-1mm high)
where water flow is slowed by frictional forces
Brackish water partially fresh and partially salt
water

Brood pouch pouch in adult in which young or eggs are carried

Bryozoa moss animals or sea mats, phylum of aquatic colonial animals

Byssal thread strong filament secreted by gland of mussel for attachment

Calcareous containing calcium carbonate (lime)

Calcicole (of plants) thriving in calcium-rich soils

Carapace dorsal covering overlying thorax of most crustaceans; often extends around side of body

Carbohydrate compound containing carbon, hydrogen and oxygen $C_x(H_2O)_y$. Essential part of metabolism in all organisms, e.g. sugar and starch

Carnivore flesh-eating organism

Carr fen woodland, usually dominated by alder or willow

Catadromous descending towards the sea to spawn

Catchment area area drained by a river and its tributaries

Chert flint-like quartz occurring in strata of chalk rocks

Chiton member of a group of molluscs whose body and mantle bear calcareous plates

Chlorophyll pigment found in green plants essential for photosynthesis

Chondrichthyes class of fish with cartilaginous endoskeleton

Cilia microscopic, hair-like, actively moving processes used in locomotion or to move substances internally

Circadian rhythm metabolic or behavioural rhythm with cycle of about 24 hours

Circuli ring-like arrangement; increments added to circumference of fish scale as animal grows

Claspers modified pelvic fins in cartilaginous fish used by male for gripping female during mating

Class taxonomic group into which a phylum is divided

Cloaca in vertebrates, terminal region of gut into which reproductive and kidney ducts open; in some invertebrates, end part of intestine (e.g. sea cucumber)

Coagulate clot, change from liquid to viscous or solid state

Cocoon protective case

Coelenterate phylum of multicellular, radially symmetrical animals, including corals, hydroids, anemones, jellyfish

Commensalism association between two animals living together and sharing food where neither animal is harmed

Conchologist person who studies and/or collects shells of molluscs

Conglomerate composed of fragments of pre-existing rocks which have been rounded and cemented together

Consumer organism which uses up organic material; cf. producer

Continental shelf shallow, submerged part of continental landmass

Copepod member of sub-class of small, free-living or parasitic crustaceans

Crèche nursery ground for young animals

Crepuscular active at dawn and dusk

Crustacean member of class of mainly aquatic, gill-breathing arthropods, which includes shrimps, crabs, water fleas, etc

Cryptic coloration camouflage against surroundings

Cuticle external, non-cellular and often impermeable layer covering an organism

Cypris larva final larval stage in barnacle life history

Cyst resistant, shell-like, proteinaceous covering of reproductive cell

DDT Dichloro-diphenyl-trichlorethane; an insecticide

Deciduous of tree, in which leaves fall at the end of growth season

Decomposer organism feeding on and breaking down dead plant and animal material

Denticle small, tooth-like process; tooth-like scales present in many sharks

Desmid unicellular or colonial green alga

Detritus particles of debris from decaying plants and animals

Diatom microscopic, single-celled alga with cell wall of silica

Diffusion movement of molecules from an area where they are more concentrated to where they are less concentrated

Dinoflagellates group of largely marine, single-celled, phytoplanktonic protozoans

Dioecious having male and female reproductive organs on separate individuals

Diurnal active during daytime; opening during daytime; occurring daily

Dorsal back or upper surface

Drift net large net extended by weights and buoys allowed to drift with current

Drumlin smooth, elongated hill composed of boulder clay and deposited by ice

Dyke channel dug to drain water, or embankment against flooding

Dystrophic lake condition rich in accumulated peat in which decomposition is inhibited by lack of calcium and nutrients in the water

Ecosystem interacting system comprising organisms and their environment

Ectoparasite parasite living on exterior of host organism

Effluent outflow (particularly of pollutant)

Elver young eel

Elytra hard, case-like outer wings of certain insects, particularly beetles

Endoparasite parasite living in interior of host organism

Endoskeleton internal skeleton

Enzyme proteinaceous substance which catalyses a biochemical reaction

Ephippium resistant case containing resting eggs of water flea

Epilimnion upper layers of water in a lake between a thermocline and the surface

Equinox one of two periods in year when day and night are of equal length

Eustatic large-scale rise or fall in sea level usually due to the formation or the melting of ice-sheets

Eutrophic nutrient-rich water with high organic production

Exoskeleton external skeleton

Fat group of organic compounds used as food reserve in animals and some plants

Filamentous thread- or strand-like

Filter-feeder animal which feeds by straining off small organisms from water

Fjord (or fiord) long, steep-sided coastal inlet formed by glacial action, and normally invaded by the sea

Flaccid limp (usually water-deficient)

Flagellum whip-like projection from cell, giving lashing or undulating movement

Flat see RIFFLE

Flatworms group of flattened, worm-like invertebrates, many of which inhabit marine and fresh waters

Flotsam floating debris at sea

Fluvioglacial deposit material deposited by melt water from glacier

Food chain chain of organisms which feed successively on each other, e.g. plant (primary producer) →herbivore →carnivore

Food pyramid loss of energy at each stage of food chain results in a large mass of primary producers supporting a smaller mass of primary consumers supporting an even smaller mass of secondary consumers

Free-living organism which moves freely and is able to survive on its own

Frond leaf-like structure

Fry young offspring; stage between alevin and parr in trout and salmon life history

Fucoid group of brown seaweeds or wracks

Fumarole hot spring emitting vapour, found in volcanic regions

Gamete reproductive cell: sperm (male), egg (female)

Gastropods group of molluscs including snails, sea slugs and sea hares

Gemma packet of cells which develops into a new organism, particularly in mosses and liverworts

Genus taxonomic group made up of closely related species

Geothermal relating to the heat of the earth's interior

Germination initial development in spore or seed

Gizzard muscular grinding chamber in gut of various animals

Glochidia larva larva of some freshwater mussels

Gonad sexual organ which produces gametes

Gorge very steep-sided, deep river valley

Gregarious tending to herd together or form clusters

Grilse salmon which returns to fresh water after only one year at sea

Groyne man-made structure built out into sea to retain beach material and reduce erosion by waves

Guano excrement of seabirds

Habitat environment in which an organism lives

Haemoglobin red respiratory pigment in blood of vertebrates and some invertebrates which has high affinity for oxygen

Halophyte plant able to tolerate and/or thrive in very salty water

Herbivore plant-eating organism

Hermaphrodite organism with both male and female reproductive organs

Holdfast disc-like attachment area of seaweed

Holt riverside den of otter

Hydrofuge water-repelling

Hypolimnion water between the thermocline and the lake bottom

Igneous rocks which have solidified from molten magma as a result of volcanic action. See also SEDIMENTARY, METAMORPHIC ROCK

Imago last, or adult, sexually mature stage of an insect

Indigenous native, not imported

Infauna burrowing animals which live in bottom sediments

Internode part of a plant stem between two successive nodes

Intertidal between extreme high and low tide marks

Invertebrate general term for all animals without a backbone

Ion atom or dissociated molecule with an electric charge

Isobar points joining places of equal pressure on a weather map, like contours

Isostatic adjustments changes in relative level of land and sea produced by the weight of ice during an Ice Age or by the melting of ice afterwards

Isotope one of two or more forms of the same chemical element which have the same atomic number but different atomic weights and slightly different physical properties

Jetsam material jettisoned to lighten load at sea (later washed ashore)

Karst scenery in hard limestone area with lack of surface drainage and numerous sink holes, caves and caverns underground

Kelt adult salmon after spawning

Kettle hole hollow left in land surface where a large ice block, left behind after the retreat of the main ice sheet, has melted

Krill general term given to food of baleen whales

Laminar flow smooth flow of water without any eddies or turbulence, which tends to form parallel layers which slide over one another with little mixing

Larva immature animal which is totally

different from adult of same species

Lateral line system of sensory organs arranged along each side of fish body which are . sensitive to vibrations

Leptocephalus larva translucent larva of eel

Lichens plant group of dual organisms formed from symbiotic associations between a fungus and an alga

Littoral intertidal part of seashore

Longshore drift process by which material is transported along the shore by wave action

Lusitanian flora and fauna found in Britain and Ireland only in the west and south-west and also in France, Spain and Portugal

Macrophyte large plant

Magma molten fluid formed in the earth's crust

Mandible lower jaw (vertebrates); mouthparts (arthropods)

Mantle specialized region of mollusc body wall which secretes shell

Mantle chamber cavity between shell with underlying mantle and the body proper of a mollusc

Mask modified, fused third pair of jaws of dragonfly nymph ending in strong hooks used for seizing prey

Meander bend in a river

Melt water run-off water produced when ice melts

Mesotrophic an intermediate state between eutrophic and oligotrophic

Metalimnion see THERMOCLINE

Metamorphic rock rock which was originally sedimentary or igneous and has undergone structural changes due to heat, pressure or natural agencies

Metamorphosis in insects, change from juvenile to adult form involving reorganization of body tissue. Complete~, life-cycle includes larval stage and wings are developed inside body; incomplete~, young are immature miniature versions of adult and wings develop outside body

Microhabitat very localized part of a general habitat in which special conditions prevail

Midden ancient, man-made waste dump

Mire swampy ground or bog

Molecule smallest portion into which a substance can be divided without losing its chemical identity

Mollusc phylum of soft-bodied and usually hard-shelled animals which include limpets, cuttle-fish, snails, octopuses and bivalves

Monocotyledon group of flowering plants with a single seed leaf in the embryo

Moraine material transported and deposited by glacier

Moult periodically to cast or shed outer covering

Mucilaginous slimy and mucus-like

Mucus slimy substance like a snail's trail

Nauplius first larval stage of certain crustaceans

Neap tide smallest ranging tide

Nematodes phylum of unsegmented worms known as 'round worms'

Nemertines phylum of mostly marine animals known as 'ribbon worms'

Neurohormone hormone secreted by neurosecretory nerve cells

Niche specific part of an ecosystem occupied by a species

Nocturnal active at night

Node position on a plant stem at which leaves arise

Nutria under-fur of coypu

Nutrient substance absorbed by a plant and used in its metabolism

Nymph juvenile, wingless form of insects with incomplete metamorphosis

Oligotrophic nutrient-poor water with low organic production

Omnivorous eating both plants and animals

Operculum flap covering gills of bony fish; horny plug used to seal some gastropod shells

Order taxonomic grouping of related organisms,

several of which form a class

Osmoregulation process of regulating concentration of body fluid

Osteichthyes class of fish with bony endoskeleton

Ostracod small crustacean having a bivalved carapace enclosing head and body

Otolith calcium carbonate granule in inner ear of fishes

Outwash plain area in front of a glacier over which melt water flows and deposits material

Ovipositor in female insects, an extension of the abdomen through which eggs are laid

Oxbow lake crescent-shaped lake left behind by meandering river

Parasitism association between two living species in which the parasite benefits at the expense of its host

Parr second-year salmon before becoming smolt

Parthenogenesis development of unfertilized egg

Perennial plants continuing growth from year to year

Periodicity regular occurrence of (some metabolic or behavioural function)

pH measure of concentration of hydrogen ions in a solution: neutral solution=7, acidic solution has pH less than 7, alkaline solution has pH greater than 7

Phaeophycin brown pigment occurring in some algae

Photosynthesis process by which green plants synthesize organic compounds from water and carbon dioxide using radiant energy

Phototaxis movement towards (positive) or away from (negative) light stimulus

Phylum major division within animal or plant kingdoms

Phytoplankton plant plankton

Plankton aquatic organisms (usually small) which live drifting in water

Pollination transfer of pollen from male parts to female parts of plant prior to fertilization

Pollutant any agent which contaminates an environment to the detriment of environmental quality

Polychaetes group of annelid (segmented) marine worms

Polyp generally sessile, and often colonial stage in life history of coelenterate

Posterior near or at the hind end

Prehensile adapted for holding

Prevailing wind wind blowing from its most commonly occurring direction

Proboscis extendible mouth structure

Producer any organism which can manufacture its own food

Proleg unjointed, leg-like appendage on abdominal surface of some fly larvae

Protein essential organic compound of animals, and plants containing carbon, nitrogen, oxygen, hydrogen and sulphur

Prothallus independent reproductive body which develops from a spore, in fern life cycle

Protozoa phylum of microscopic, single-celled organisms, comprising both plant- and animal-like species

Protrusible capable of protruding, extensible

Pseudopod foot-like projection

Pupa in insects with complete metamorphosis, a stage where insect is enclosed and much tissue reorganization occurs before adult emerges

Purse seine net fishing net shot out around shoal and then bottom enclosed by pursing with draw-rope

Quadrat frame within which animal and plant communities are sampled

Radula horny, teeth-bearing 'tongue' of molluscs

Regurgitate bring food material back up again from stomach

Relative humidity amount of atmospheric moisture relative to the maximum possible for a given temperature

Rheotaxis movement towards (positive) or away

from (negative) current stimulus (usually water)

Rhizoid uni- or multi-cellular thread-like outgrowth serving as root (in lower plants)

Rhizome horizontal underground stem

Ria river valley submerged by sea to become a long, winding inlet

Ribbon worm see NEMERTINES

Riffles and flats riffle: fast-flowing shallow, shingly area of river or stream bed adjacent to flat: slower-flowing, deeper area with finer sediment

Rotifers phylum of microscopic aquatic, ciliated, multicellular animals

Salinity degree of saltiness expressed in parts per thousand (°/ₒₒ)

Saturated solution solution which has dissolved as much of a compound as it can

Scavenger animal feeding on carrion

Sea squirts common name for sub-phylum of small, marine animals

Sedimentary formed by the deposition of sediments

Seismic profile recordings obtained from earthquakes or from explosions of the deep structure of rocks

Sessile living attached to substrate (rock, shell or seaweed)

Seta(e) bristle-like structure(s)

Sexual dimorphism marked differences between male and female of same species

Sinusoidal movement eel-or snake-like undulations of the body

Siphon tube-like structure for passage of water or other fluids

Siphonophore order of colonial coelenterates

Slacks low-lying marshy areas between coastal dunes

Smolt salmon two or more years old

Snout front portion of glacier

Sonograph picture produced by reflection of sound waves

Spat spawn or young of bivalve molluscs

Species smallest, commonly used taxonomic grouping

Spermatozoid free-swimming male gamete in lower plants

Spiracle respiratory opening

Spit extension from coast of beach-like material, deposited by longshore drift, which curves landwards at its free end

Splash zone region of shore above high-water mark subject to sea spray

Spore small, usually unicellular, resistant reproductive body produced by plants, protozoa and bacteria

Sporling young plant developing from spore

Spring tide biggest ranging tide

Stack remnant of headland which becomes detached as arch collapses

Statoblast specialized bud or resistant egg of some bryozoans

Stipe stalk, as in seaweed or fungi

Stipule small, leaf-like process on either side of leaf-stalk in many plants

Stoma(ta) pore(s) in epidermis of plants

Storage capacity amount of water the ground can hold (increases with permeability of rock)

Strandline line of debris left at high water mark when tide recedes

Strata bed of sedimentary rock usually consisting of a series of layers of the same kind

Style female part of flower which receives the pollen

Subimago stage between pupa and imago of some insects

Sublittoral zone immediately below low water spring tide

Suspension feeder feeding on organisms of detritus suspended in surrounding water

Swallow hole funnel-shaped cavity in limestone which permits passage of water underground, created by rain water dissolving the rock

Symbiosis association between two species in which one or both gain from the other's presence and neither is harmed.

187

Tap root vertical, tapering main root from which lateral roots project

Tertiary bed rock strata laid down during Eocene to Pliocene periods (relatively recent in terms of geological time)

Test shell or hardened outer covering

Thallus simple plant body not differentiated into stem and leaf

Thermocline water layer of rapidly changing temperature; metalimnion

Thorax region between head and abdomen of arthropods

Thyroid gland endocrine gland in neck of vertebrates, secreting hormone

Till boulder clay

Tombolo spit or bar which joins an island either to the mainland or to another island

Topography geographical features of a region or locality collectively, both natural and artificial

Tributary side-stream or river flowing into a main river

Trolling type of sport fishing using swiftly towed line

Tun barrel-shaped form of water bear or tardigrade, allowing it to survive prolonged drought

Turbidity degree of cloudiness of water due to suspended matter

Vegetative portion fragment of plant which splits or breaks from parent plant and develops

Vegetative reproduction the formation of other individuals without sexual reproduction; in animals by budding or splitting, in plants by underground tubers, rhizomes etc.

Ventral front or lower surface

Voe bay, creek or inlet

Water table level to which groundwater builds up above an impermeable rock layer

Wave-cut platform flat area of solid rock cut by wave action and exposed between high and low tide

Wrack brown seaweed

Xerophyte plant adapted to very dry conditions

Zone band on shore in which certain species predominate; or clearly defined area distinguished from adjacent areas by certain climatic or topographical conditions

Zooplankton animal plankton

Zooxanthellae yellow-brown, symbiotic, unicellular algae

BIBLIOGRAPHY

★ Field guides and keys ○ Estuarine
● Marine □ General books
△ Freshwater ◆ Specific locations

○ Angel, Heather 1974 *The World of an Estuary*. Evans, London.

● Angel, Heather 1975 *Photographing Nature: Seashore*. Fountain Press, Argus Books, King's Langley.

★ Bang, P. and Dahlstrom, P. 1974 *Collins Guide to Animal Tracks and Signs*. Collins, London.

●○ Barnes, R. S. K. (ed) 1977 *The Coastline*. Wiley, Chichester and Toronto.

●○ Barnes, R. S. K. 1979 *Coasts and Estuaries*, ed. J. Ferguson-Lees and B. Campbell. Hodder and Stoughton, London.

★● Barrett, J. and Yonge, C. M. 1972 (rev. edn) *Collins Pocket Guide to the Sea Shore*. Collins, London.

★△ Belcher, H. and Swale, E. 1976 *A Beginner's Guide to Freshwater Algae*. HMSO, London.

★△ Belcher, H. and Swale, E. 1979 *An Illustrated Guide to River Phytoplankton*. HMSO, London.

□ Buchsbaum, R. 1953 *Animals Without Backbones* (2 vols). Penguin Books, Harmondsworth; 1975 University of Chicago Press.

●Campbell, A. C. 1976 *The Hamlyn Guide to the Seashore and Shallow Seas of Britain and Europe*. Hamlyn, London.

★△ Clegg, J. 1965 (3rd edn) *The Freshwater Life of the British Isles*. Warne, London.

△ Corbett, P., Longfield, C. and Moore, N. W. 1960 *Dragonflies*. Collins, London.

★● Cramp, S., Bourne, W. R. P. and Saunders, D. 1975 (2nd rev. edn) *The Sea Birds of Britain and Ireland*. Collins, London.

★ Darlington, A. and Smyth, J. C. 1967 (rev. edn) *Keys to Small Organisms in Soil, Litter and Water Troughs*. Longman, London, and Penguin Books, Harmondsworth.

□ Deacon, G. E. R. (ed) 1962 *Oceans*. Hamlyn, London.

□● Dickinson, C. I. 1963 *British Seaweeds*. The Kew Series, Eyre & Spottiswoode, London.

△ *Directory of Wetlands of International Importance in the Western Palearctic*. Compiled by Erik Camp for UNEP and IUCN, publ. UNEP and WWF Gland, Switzerland.

● Eddison, J. 1979 *The World of the Changing Coastline*. Faber & Faber, London.

● Eltringham, S. K. 1971 *Life in Mud and Sand*. English Universities Press, London.

□ Falkus, H. 1978 *Nature Detective*. Gollancz, London.

★● George, D. and George, J. 1979 *Marine Life, An Illustrated Encyclopedia of Invertebrates in the Sea*. Harrap, London.

★Gooders, John 1970 *Where to Watch Birds in Britain and Europe*. Andre Deutsch, London.

○ Green, J. 1968 *The Biology of Estuarine Animals*. Sidgwick and Jackson, London, and University of Washington Press.

○● Hale, W. G. 1980 *Waders*. Collins, London.

● Hardy, A. C. 1970 (rev. edn) *The Open Sea: Part 2. Fish and Fisheries*. New Naturalist Series, Collins, London.

● Hardy, A. C. 1971 (rev. edn) *The Open Sea: Part 1. The World of Plankton*. New Naturalist Series, Collins, London.

◆ Harrison, J. and Grant, P. 1976 *The Thames Transformed, London's River and its Waterfowl*. Andre Deutsch, London.

● Harrison, R. J. and King, J. E. 1980 (2nd edn) *Marine Mammals*. Hutchinson, London.

★△ Haslam, S. M., Sinker, C. A., and Wolseley, P. A. 1975 'British Water Plants', reprinted from *Field Studies* 4:243–351.

△ Haslam, S. M. 1978 *River Plants*. Cambridge University Press, Cambridge.

★△ Haslam, S. M. and Wolseley, P. A. 1981 *River Vegetation*. Cambridge University Press, Cambridge.

□● Hepburn, I. 1952 *Flowers of the Coast*. Collins, London.

□ Hickling, C. F. 1975 *Water as a Productive Environment*. Croom Helm, London.

□ Hynes, H. B. N. 1960 *The Biology of Polluted Waters*. Liverpool University Press, Liverpool.

◆△ Jónasson, P. M. (ed) 1980 *Lake Mývatn*. Icelandic Literature Society, Copenhagen.

□△ Jonsson, Lars 1977 *Birds of Lake, River, Marsh and Field*. Penguin Books, Harmondsworth.

● Lewis, J. R. 1964 *The Ecology of Rocky Shores*. English Universities Press, London.

□△ Loffler, H. 1974 *Der Neusiedlersee*. Fritz Molden, Vienna.

★● Luther, W. and Fiedler, K. 1976 *A Field Guide to the Mediterranean Sea Shore*. Collins, London.

★△ Macan, T. T. 1959–66 *A Guide to Freshwater Invertebrate Animals*. Longman, London.

△ Macan, T. T. 1973 *Ponds and Lakes*. George Allen & Unwin, London; 1974 Crane-Russak.

△Macan, T. T. and Worthington, E. B. 1951 (paperback 1972) *Life in Lakes and Rivers*. New Naturalist Series, Collins, London.

○ McLusky, D. S. 1981 *The Estuarine Ecosystem*. Blackie, Glasgow.

□★ McMillan, N. F. 1968 *British Shells*. Warne, London.

★△ Mellanby, H. 1963 (paperback 1975) *Animal Life in Fresh Water*. Methuen, London, and Halsted Press, Boston.

□ Money, D. C. 1970 *The Earth's Surface*. Evans, London.

△ Moss, B. 1980 *Ecology of Freshwaters*. Blackwell Scientific Publications, Oxford and Boston.

★△ Muus, B. J. and Dahlstrom, P. 1971 *Collins Guide to the Freshwater Fishes of Britain and Europe*. Collins, London.

★● Muus, B. J. and Dahlstrom, P. 1974 *Collins Guide to the Sea Fishes of Britain and North-western Europe*. Collins, London.

● Nelson, B. 1980 *Seabirds, their Biology and Ecology*. Hamlyn, London.

● Newell, R. C. 1970 *Biology of Intertidal Animals*. Elek, London.

★● Newell, G. E. and Newell, R. C. 1973 (rev. edn) *Marine Plankton*. Hutchinson, London.

● Nicol, J. A. 1967 (2nd edn) *The Biology of Marine Animals*. Pitman, London.

□△ Ogilvie, M. A. 1979 *The Bird-Watcher's Guide to the Wetlands of Britain*. Batsford, London.

★△ Pott, E. (trans. G. Vevers) 1980 *Rivers and Lakes*. Chatto and Windus, London.

● Russell, F. S. and Yonge, C. M. 1975 (rev. edn) *The Seas*. Warne, London.

□ Smith, M. 1951 *The British Amphibians and Reptiles*. Collins, London.

● Soper, T. 1972 *The Shell Book of Beachcombing*. David and Charles, Newton Abbot.

● Soper, T. 1979 *Beside the Sea*. BBC Publications, London.

● Steers, J. A. 1972 *The Sea Coast*. Collins, London.

□ Tansley, A. G. 1968 (2nd edn rev. M. C. F. Proctor) *Britain's Green Mantle*. Allen and Unwin, London.

★● Tebble, N. 1976 (2nd edn) *British Bivalve Seashells*. British Museum (Natural History) Publications, London.

□ *Waterways and Wetlands* 1976. Compiled by Alan Brooks for British Trust for Conservation Volunteers, London.

◆ Weber, K. and Hoffmann, L. 1980 *Camargue*. Kümmerly and Frey, Berne.

□ Wheeler, A. 1969 *The Fishes of the British Isles and North West Europe*. Macmillan, London.

◆ Wheeler, A. 1979 *The Tidal Thames: History of a River and its Fishes*. Routledge & Kegan Paul, London.

△ Whitton, B. A. 1975 *River Ecology*. University of California Press.

△ Whitton, B. A. 1979 *Rivers, Lakes and Marshes* ed. J. Ferguson-Lees and B. Campbell. Hodder and Stoughton, London.

●Yonge, C. M. 1949 *The Sea Shore*. New Naturalist Series, Collins, London.

INDEX

191